THE DIPLOMACY OF PRAGMATISM

American Diplomatic History
Lawrence S. Kaplan, Editor

Aftermath of War: Americans and the Remaking of Japan, 1945–1952
Howard B. Schonberger

The Twilight of Amateur Diplomacy: The American Foreign Service and Its Senior Officers in the 1890s
Henry E. Mattox

Requiem for Revolution: The United States and Brazil, 1961–1969
Ruth Leacock

American Historians and the Atlantic Alliance
edited by Lawrence S. Kaplan

The Diplomacy of Pragmatism: Britain and the Formation of NATO, 1942–1949
John Baylis

The Diplomacy of Pragmatism

Britain and the Formation of NATO, 1942–1949

John Baylis

The Kent State University Press
Kent, Ohio

Published in the United States by
The Kent State University Press
Kent, Ohio 44242

Printed in Hong Kong

Library of Congress Catalog Card No. 92–5602

ISBN 0–87338–471–7

Library of Congress Cataloging-in-Publication Data
Baylis, John
The diplomacy of pragmatism: Britain and the formation of NATO, 1942–1949 / John Baylis.
p. cm. — (American diplomatic history)
Includes bibliographical references and index.
ISBN 0–87338–471–7 (cloth: alk. paper) ∞
1. North Atlantic Treaty Organization—History. 2. North Atlantic Treaty Organization—Great Britain—History. 3. Great Britain—Foreign relations—1945– 4. Europe—National security. 5. Great Britain—National security. I. Title. II. Series.
D845.B37 1993
355'.031'091821—dc20 92–5602
CIP

To Brian

Contents

Acknowledgements

I would like to express my thanks to the British Academy for a grant in 1983 which enabled me to do research at the National Archives in the United States. I am particularly grateful to Professor Inis Claude and the Department of Government and Foreign Affairs for giving me a 'home' at the University of Virginia, Charlottesville during this research trip to the United States. My thanks also go to the University of Wales, Aberystwyth for granting me sabbatical leave to complete the research on the book and for providing financial support to visit the Canadian Public Archives in September 1990. I am also grateful to Professor John Garnett for providing me with additional financial support from the Wilson Fund to visit Canada.

I also wish to record my gratitude to Lord Gladwyn for allowing me an interview and to Sir Nicholas Henderson for the considerable trouble he has gone to in corresponding with the author on various occasions. Dr Cees Wiebes, Dr Bert Zeeman, Dr Peter Foot, Dr John Young and Dr Alex Danchev have also provided very useful advice on various topics covered in the book.

I am grateful to the library staffs of numerous institutions and archives for their assistance. In particular I wish to express my thanks to the staff of the George C. Marshall Library, at Virginia Military Institute; the Library of Congress; the National Archives in Washington; the Harry S. Truman Library in Independence; the Public Archives of Canada in Ottawa; Churchill College, Cambridge; and the British Library of Political and Economic Science. I am especially grateful to the staff at the Hugh Owen Library in Aberystwyth for their cheerful assistance and the staff at the Public Record Office in London. Copyright material from the PRO has been used with the permission of the Controller of Her Majesty's Stationery Office.

Some of the material in this book originally appeared in *International Affairs* and *The Journal of Strategic Studies*. I am grateful to the editorial boards of both journals for allowing me to use the material in this study.

My thanks also go to Carol Parry who has borne the arduous task of typing the manuscript for this book with great cheerfulness and skill. I am also grateful to Alan Macmillan who, for the third time in recent years, has undertaken the tedious task of indexing a book for me.

Finally, and most importantly, I am indebted to my wife, Marion, and daughters, Emma and Katie for their moral support and encouragement, and to my brother, Brian, for his considerable hospitality and good counsel over the years.

Preface

In the years 1989 and 1990 the whole face of European security was transformed. Initially, people's revolutions took place all over Eastern Europe, dramatically sweeping away the reactionary Communist regimes which had been in power since the 1940s. This was followed by free elections and an uneven process of democratic reform. Unlike previous challenges to Communist rule, there was no intervention by the Soviet Union, which was under the reformist leadership of Mikhail Gorbachev. Indeed the Soviet leader indirectly encouraged the process of reform and continued to pursue a wide range of remarkable foreign policy initiatives designed to transform East West relations. Unilateral cuts in defence expenditure, the withdrawal of Soviet troops from Afghanistan and Eastern European countries, unprecedented concessions in arms control negotiations, all signalled a fresh approach to diplomacy and security. With the signing of the Treaty on the Final Settlement with Respect to Germany by the four allied states (Britain, the United States, the Soviet Union and France) and the two Germanies on 12 September 1990 and German reunification on 3 October of the same year, the forty-five year division of Germany was ended. 1991 also saw the disintegration of the Soviet Union. To most contemporary observers of the postwar era the Cold War finally had been brought to a close.

However, these breathtaking events brought uncertainties over future European security arrangements. As a result, one of the key tasks for statesmen in the 1990s is the creation of a new European security order. Despite its flaws the old postwar security system based on confrontation and nuclear deterrence had helped to keep the peace for an unprecedented period in European history. Now that the Cold War has ended and the foundations of the old order have been irretrievably undermined, the key question is what is going to take its place. A new European order has to be built, different from the old, taking account of the new superpower relationship and the rapidity of the changes taking place in the Western and Eastern halves of Europe.

In this context, in which uncertainty prevails and new turbulent forces are beginning to stir, the history of the formation of the North Atlantic Treaty Organisation in the late 1940s is of some interest.

In some important respects the world of the 1990s and the world of the late 1940s are very different. In particular, Britain's status in the international community is very different. In the late forties a major hot war had just ended and the victorious allies had to struggle with the reemergence of strong ideological suspicions and conflicting interests. In the 1990s, although the Cold War has just ended and some residual suspicions remain, a much greater consensus is evident among the leading powers on the need to build a system of security which is less dependent on the kind of military confrontation and division of Europe which has characterised the last forty-five years. In particular, ideological confrontation between the major powers is less pronounced and the former Soviet Republics are much more concerned with the process of domestic reform than with achieving a *cordon sanitaire* in Eastern Europe. At the same time democracy is breaking out all over Eastern Europe. Some of the structures of the old system of security also remain in place to help provide confidence as the process of transformation is taking place. Unlike the late 1940s Germany is also now reunited under a democratic government. In some respects, therefore, the international environment of the early 1990s appears much more auspicious for the continuation of global harmony compared with the gathering clouds of the immediate postwar years. In other respects, the withdrawal of Soviet forces from the Eastern European states, the disintegration of the Soviet Union, and the civil war in Yugoslavia have raised new problems which are very different from those of the late 1940s.

Despite these important differences, however, there are some similarities between the two periods which makes the story of the formation of NATO instructive. In the late 1940s the old classical system of security had been totally undermined by the events of the Second World War. By 1990 the foundations of the postwar order had also been significantly eroded. Statesmen in the late 1940s were faced with great uncertainties as they attempted to develop new concepts and structures designed to establish a more stable and peaceful international order. Much the same is true in the early 1990s. In the late 1940s the first priority was to build a system based on cooperation with the Soviet Union. The need to establish cooperative relations between the West and the former Soviet Republics remains essential in the present era.

In the late 1940s a new order only emerged as a result of a combination of pragmatism, patience and vision. In this respect the history of the formation of NATO is particularly illuminating. Immediate problems had to be resolved. Flexible and painstaking diplomacy was necessary. New concepts had to be forged. A creative process of developing a new security system had to be undertaken. Keeping different options open, trying continuously to blend different economic, political and security dimensions of policy together and keeping sight of the general direction it was hoped to move in, were all characteristics of British policy in the years which led to the signing of the North Atlantic Treaty.

In the 1990s the same kind of blend of pragmatism, patience and vision is necessary if a new European order is to be established. A wide variety of contemporary structures compete for preeminence in the search for a new security framework. At the same time security depends to an even greater extent than in the past on major economic and political changes which are taking place in the world in general and Europe in particular. In this uncertain environment a single concept or structure is unlikely to be adequate as the basis of the new European order. A patient search for different institutions to resolve different problems is more likely to be successful. The story of Britain's role in the formation of NATO provides an interesting illustration of how pragmatic diplomacy helped to lay the foundations of the postwar system of European security. It also provides a warning that a single, predominantly military structure of security is only second-best. Lasting security depends on combining a wide range of economic and political relationships as well as military power. Equally, without the confidence that only military security can provide, the development of a broad conception of security is likely to prove difficult to achieve.

Introduction

This book is not directly concerned with the origins of the Cold War. In dealing with Britain's role in the formation of the North Atlantic Treaty Organisation, however, it is inevitably set in the context of the evolution of Western policies in the early stages of the Cold War. It is important, therefore, to explain at the outset the position adopted by the author in the debates which have taken place in the literature about the origins of the Cold War.

It has been argued that in terms of Cold War historiography the 1950s were dominated by orthodoxy or traditionalism, the 1960s by revisionism and the 1970s by post-revisionism.[1] Traditionalist historians like W. H. McNeill and Hugh Seton-Watson laid the blame for the Cold War squarely on the shoulders of the Soviet Union.[2] According to this school it was Stalin's aggressive pursuit of worldwide domination which forced the United States to react defensively and to organise the Western powers into an effective alliance to contain Soviet ambitions.

In the 1960s, in the era of American involvement in Vietnam, this view was challenged by writers such as William Appleman Williams and Gar Alperowitz.[3] Their revisionist thesis emphasised American economic imperialism as being the main cause of the Cold War. In their view it was the determination of the United States to expand its economic influence and its predilection for using atomic diplomacy, which forced Stalin to act defensively and consolidate Soviet control, particularly in Eastern Europe.

Revisionism was followed by post-revisionism in the 1970s when writers like John Lewis Gaddis tried to provide a more balanced analysis.[4] In Gaddis's view both the United States and the Soviet Union shared responsibility for the Cold War. The Soviet Union was a little more to blame than the United States, because of Stalin's aggressive policies in Eastern Europe, but American economic imperialism is regarded as an important contributory factor in the breakdown in relations between the two countries.

The views of the post-revisionists represented a major step forward in Cold War historiography in bringing together the most convincing arguments from the traditionalist and revisionist schools of thought. Post-revisionism itself, however, was subject to major criticism in the early 1980s by writers like Warren Kimball and Bert Zeeman.[5] Kimball argued that despite Gaddis's attempt to combine the best features of traditionalism and revisionism he failed to produce 'a new thesis or synthesis' which had any real substance.[6] The main problem with post-revisionism, he argued, was that it had no 'new philosophical content' of its own.[7]

This is an argument which has been taken up by Bert Zeeman. In his view, 'by disassociating himself from revisionism Gaddis is led back into the arms of orthodoxy; and to encounter the inescapable reproach that post-revisionism is simply orthodoxy plus archives, Gaddis has to creep back to revisionism by confessing that the concept of empire (albeit by invitation) is the central feature that post-revisionism has inherited from revisionism'.[8] According to Zeeman, Gaddis's dilemma is insoluble.

There is another criticism of post-revisionist historians which applies equally to traditionalists and revisionists. Most of the writing on the subject until recently has tended to be by American historians and, irrespective of whether they were traditionalists, revisionists or post-revisionists, their approach is rather parochial. Their focus is on the American role in the Cold War. As a result the Cold War is invariably seen as a bi-polar confrontation between the United States and the Soviet Union in which the role of other states is marginalised. Britain, France, Germany and the other European states tend to be discussed only to the extent that they impinge on 'the Great Conflict'.

A significant attempt was made to redress this over-narrow view of the origins of the Cold War in an open letter written by Donald Cameron Watt in *The Political Quarterly* in 1978.[9] In his letter Watt pointed to the opportunity created for historians by the opening of the British public records for the postwar period. These new archives, he argued, would force historians to confront important questions about Cold War historiography. They had an important choice to make. They could either use the information becoming available to add to the debates taking place in American historical circles over who should be blamed for the start of the Cold War, or they could address themselves 'to the true task of historians' of understanding and elucidating the 'causes and course of what we have come to think of as the Cold War'.[10]

Watt had little doubt about which course historians should follow. In his view the traditionalist–revisionist debate was unhistorical because it tried to apportion blame and because 'the line of argument was from the present to the past instead of the other way around'.[11] The whole debate about the Cold War had become embroiled with the ideological disputes of the 1960s and 1970s which coloured and distorted the historical judgements of the 1940s. The bi-polar framework of the later period had been imposed on the immediate postwar events.

Watt urged historians to make a fresh start and not to become involved in the unhistorical debates about the responsibility for the Cold War. It was important, he argued, 'to behave as historians' and not to 'hunt down les coupables' in the American way.[12]

This was sound advice and appears to have had a considerable impact on historians, at least in Western Europe, in recent years. The result has been a growing literature based on wide-ranging archival research which has begun to break down the narrow bipolar views of the Cold War. The works of Robert Hathaway, Terry Anderson, Bert Zeeman and Cees Wiebes, Nicholaj Petersen, Elizabeth Barker, John Young, William Roger Louis, William Cromwell, James Gormly, Victor Rothwell and Sean Greenwood are all notable in this respect.[13] Robert Hathaway has summed up this new approach to the origins of the Cold War as one of 'depolarization'.[14] The Cold War 'is no longer seen simply as a conflict between the United States and the Soviet Union'.[15] In contrast, as Bert Zeeman has argued, the new contributions stress the fact that Britain, in particular, 'more often than had been acknowledged by traditionalists, revisionists and post-revisionists alike, took the lead in the formation and implementation of a containment policy towards the Soviet Union'.[16] This emphasis on the distinctive and leading roles of other countries apart from the United States and the Soviet Union in the origins of the Cold War represents an important step forward in the literature of the Cold War. Escott Reid's book *Time of Fear and Hope* on Canada's role in the formation of NATO is a particularly good example of this broadening of the focus of writing on the subject.[17]

In the same way that traditionalism, revisionism and post-revisionism are products of the 1950s, 1960s and 1970s, it must be acknowledged that 'depolarization' is also the product of the 1980s and 1990s. As a school of thought it reflects the gradual decline of the superpowers and the emerging multipolar structure of power in international relations. In terms of European historians, no doubt, it also reflects the growing confidence and independence of the European states. It is no less useful for this, however, in the methodological improvements it brings and the broader, more accurate, interpretation it provides of the origins of the Cold War.

This book, in focusing on Britain's role in the formation of the NATO alliance, reflects this emphasis on 'depolarization'. The aim is to chart the evolution of British planning and policy towards what was initially a Western European group and eventually, from 1941, when the idea was first mooted, an Atlantic security framework to the signing of the North Atlantic Treaty on 4 April 1949. The approach which is adopted, however, attempts to avoid being overly chauvinistic. The key role of other states in the formation of NATO is acknowledged. Britain certainly did not establish NATO single-handed. Nor was British diplomacy wholly consistent or completely successful throughout the period covered. Different strands of policy struggled for preeminence. This said, the focus of the book is on Britain and the distinctive role she played in reacting to the perceived challenge from the Soviet Union from 1944 to 1949. In particular, emphasis is given to the contribution of the Foreign Secretary, Ernest Bevin, who, it is argued, combined a pragmatism and vision in coordinating the Western European states and 'entangling' the United States in the security affairs of Western Europe. In this one important area of the Cold War British initiatives were decisive. An attempt is also made to set NATO in the context of the broader strategic planning which took place in Britain at the time.

The first chapter deals with the evolution of British thinking about postwar European security from Trygve Lie's initiative in November 1940, urging Britain and the United States to establish bases in Norway after the war ended, to the detailed planning which took place in 1944 for a postwar Western European security grouping. The main purpose of the chapter is to show that ideas about European security did not emerge miraculously from nowhere in 1945. They had their roots in the years that went before.

Chapter 2 deals with the differences which emerged in 1944 and 1945 between the Foreign Office and the Chiefs of Staff over whether to plan for the possibility that great-power cooperation might break down in the postwar period. Although by the end of the war the Foreign Office and the COS agreed on some issues of postwar planning important differences remained over how to deal with the Soviet Union and what was the most appropriate framework for security in the postwar world.

Chapter 3 deals with the evolution of British attitudes towards the Soviet Union in the immediate postwar period. The chapter shows that despite the growing consensus between the Foreign Office and the Chiefs of Staff on the aggressive nature of Soviet policy the Foreign Secretary, Ernest Bevin, continued to search for accommodation with the Soviet leaders. It was in this context of deteriorating relations and uncertainty about Britain's future role that Bevin fashioned his policies towards Western Europe.

Chapter 4 deals with the evolution of Britain's relations with France from 1945 to March 1947. Ernest Bevin came to the Foreign Office with his own ideas of greater European cooperation. These coincided with Foreign Office plans and in August 1945 the Foreign Secretary declared his interest in achieving greater coordination in economic, political and military terms with Western Europe, starting with France. The chapter charts the immediate difficulties which had to be overcome and the events which led to the signing of the Dunkirk Treaty in March 1947. This is portrayed as a product of both expediency and vision in the conduct of British diplomacy.

The fifth chapter considers the gradual changes which took place in British foreign policy between March 1947 and March 1948. This was a period of major transition in British policy as the Cold War gathered momentum. Bevin still struggled to maintain cooperation with the Soviet Union but these efforts finally collapsed with the failure of the Foreign Ministers Conference in London in December 1947. Bevin's key objective during this period seems to have been to establish a 'Third Power' in world politics by fostering greater cooperation in Western Europe, maintaining Britain's position in the Middle East, and developing closer links with the Dominions and African colonies. Bevin hoped to organise the 'middle of the planet' into a coherent grouping which would reestablish Britain's power in the world. In the short term this required American economic and military support but the aim in the longer term was a more independent role. The launching of the Western Union in January 1948 and the establishment of the Brussels Pact reflected the pursuit of Bevin's wider objectives.

Chapter 6 steps back from the evolution of British foreign policy towards Western Europe and focuses on the debates taking place within the Chiefs of Staff Committee on the direction of British defence policy. It is suggested that in order to understand the attitudes of the Service Chiefs towards a continental commitment and close military ties with the United States it is important to consider the economic context in which defence planning was taking place, the interest which the CoS had in establishing a Commonwealth defence arrangement and the priority given to the Middle East in British strategy. These preoccupations, together with a much clearer and less ambiguous perception of the Soviet threat than existed in the Foreign Office, led to a reluctance by the Service Chiefs to commit forces to the defence of Western Europe. The Chief of the Imperial General Staff, Bernard Montgomery, was the exception amongst the Chiefs of Staff and during 1948 he waged an almost single-handed campaign to persuade the Chief of the Air Staff, Lord Tedder and the Chief of the Naval Staff, Sir John Cunningham, that the traditional emphasis on a maritime (or more recently maritime/air) strategy could not be sustained given the new focus in British foreign policy on Western Europe. This was a campaign which was to be pursued by Montgomery's successor, General Slim. The chapter shows that, although a limited commitment was agreed in May 1948, it took two years (until March 1950) before a fuller military commitment was made to continental defence. This reluctance by the CoS to accept a continental commitment was largely due to their belief, consistently expressed from 1944 onwards, that Western Europe was not strong enough to provide for its own defence, together with their concern to maintain a global 'Commonwealth' strategy. The CoS believed that only the military capability and resources of the United States could match the power of the Soviet Union. In particular, they were reluctant to support the Foreign Office attempt to play a leading role in the coordination of Western European defences. Political guarantees to Western Europe were meaningless, the CoS felt, without the ability to provide material military support. Given Britain's economic difficulties and the priority of the Middle East in British strategic planning, the CoS argued that Britain could only accept a continental commitment if the United States were prepared to play a part in the defence of Western Europe. In order to secure such American support, as the Cold War gathered momentum, the Foreign Office gradually amended its 'grand design' based on the Western Union, to accept a more dependent relationship with the United States in an Atlantic Alliance advocated by the Chiefs of Staff.

Having emphasised the contribution of the Service Chiefs to the Atlantic strand in British policy, attention returns in Chapter 7 to the 'Pentagon Talks' of March 1948 which reflected the growing importance of Atlanticism even in the Foreign Office. Close relations with the United States had always been accepted in diplomatic circles but, from the end of December 1947 onwards, involving the United States directly in Western European security became an important objective of British foreign policy. The secret 'Pentagon Talks' between Britain, the United States and Canada were of crucial importance in laying the foundations of the North Atlantic Treaty Organisation. They provided an indispensable formula which, unseen, guided future negotiations.

As Chapters 8 and 9 show, however, difficulties still remained in actually achieving an American commitment to the defence of Western Europe. The political sensitivities associated with an election year in the United States meant that when negotiations got under way belatedly, between the United States, Canada and the five Brussels Pact Powers, it took some time before the Americans were prepared to accept a formal treaty arrangement. Even when a draft treaty was agreed and the American election was over, problems still arose over a variety of issues right through until March 1949. The chapters chart the evolution of the negotiations in Washington through to their successful conclusion.

The signing of the North Atlantic Treaty on 4 April 1949 finally provided Britain with a solid framework upon which to base its foreign and security policies after the uncertainties of the immediate postwar period and the earlier ideas of a more independent Western Union. In the conclusion an attempt is made to set NATO in the broader context of Britain's global strategic planning. The Middle East had been the geographic centre of Britain's global strategy up to 1949. With the formation of NATO and the subsequent decision to concentrate on a continental commitment the balance in Britain's 'Commonwealth defence policy' shifted more to the

Atlantic and Western Europe. This helped to achieve a greater harmony between British foreign and defence policy. Until mid-1948 Bevin had attempted to keep his options open and retain an independent role for Britain. Despite strained relations with the Soviet Union, continuing attempts were made to reach an accommodation. At the same time, attempts were made to secure close relations with the United States; to play a leading role in Europe; to coordinate the Commonwealth; and to maintain Britain's traditional position in the Middle East. The long-term objective appears to have been to reestablish Britain's power in the world. For military planners the objectives were more straightforward: to achieve the strategic unity of the Commonwealth (broadly defined) in close alliance with the United States to meet what was regarded as an unambiguous threat from the Soviet Union to British interests world-wide. The formation of NATO helped to bring these two lines of policy more effectively into line. However, it also represented a compromise between the CoS concern to maintain close relations with the United States while remaining aloof from Europe and the Foreign Office concern to play an important part in coordinating the policies of Western European states.

The chapter concludes with an assessment of Bevin's role in the formation of NATO. Although Bevin's policies were not wholly successful and a number of criticisms can legitimately be levelled against his conduct of foreign affairs, NATO was nevertheless his greatest achievement. In some ways it reflected the failure of his earlier policies of trying to maintain cordial relations with the Soviet Union and his ambitions of achieving a 'third power' role. Nevertheless, this should not detract from the success of the patient and effective diplomacy which he and his officials pursued for much of 1948 and 1949 in the negotiations that led to the formation of the Atlantic Alliance. NATO may not have been the ideal solution but it played a key role in reinforcing British interests during the Cold War era.

1 Wartime Planning for a Postwar European Security Group, 1941–44

During the 1920s and 1930s Britain retained a close interest in European affairs but consistently steered away from any form of major military involvement on the continent. In traditional terms policymakers continued to recognise the importance of maintaining a favourable balance of power in Europe and preventing any hostile state from dominating the Low Countries. At the same time there was a fundamental political assumption that Britain and France would be in alliance in the event of renewed German aggression in Western Europe. Until Munich, however, the emphasis in defence planning centred on home and imperial defence. When Sir Thomas Inskip, the Minister for the Coordination of Defence, presented a list of defence priorities to his Cabinet colleagues in December 1937 the continental commitment was ranked last.[1] In a war with Germany, it was decided, Britain's contribution would consist mainly of naval and air forces with a very small expeditionary force.

By March 1939 events in Europe caused a major reappraisal in the thinking of the Chiefs of Staff.[2] The Service Chiefs now recognised, somewhat belatedly, that Britain's security was inextricably linked with that of France. This meant that Britain must be prepared to participate in 'the last defence of French territory'.[3] The conversion from a 'limited liability' towards European security to a continental commitment, however, came too late as German armies marched across Europe to the Channel and the remnants of the British Expeditionary Force had to be evacuated from Dunkirk in May and June 1940.

The outbreak of war and the early military reverses inevitably produced a questioning in Britain of past military policies and recommendations for new security arrangements for the postwar period. The idea of a postwar Atlantic security system was first raised by Trygve Lie with British Foreign Office officials in November 1940 just before he took up his post as Acting Foreign Minister in the Norwegian government-in-exile in London.[4] Lie suggested that in the postwar period Britain and the United States should set up bases in Norway, with all three countries maintaining outposts in Greenland, Iceland and the Faroes. These ideas were received favourably in the

Foreign Office. In April 1941 Sir Orme Sargent, Assistant Under-Secretary at the Foreign Office, wrote that

> one of the major post-war problems will be to enable this country to maintain its position vis a vis the Continent of Europe, and it is fairly evident that the failure of France will render the cooperation of the United States essential for this purpose. May not the extension of M. Lie's ideas offer a practical means of achieving this cooperation? Just as there would be, according to M. Lie's suggestion, British and Norwegian bases in Norway, there might be similar Anglo-American bases in Irish ports and even in British ports.[5]

Although Lie's initiative helped to encourage the Foreign Office to begin to think about the requirements of British defence policy in the postwar world it took some time before the bases plan was considered seriously. In early 1941 the Ministerial Committee on Reconstruction Problems was set up to look into 'practical schemes of reconstruction, to which effect can be given in a period of . . . three years after the war' and 'a scheme for a post-war European and world system'.[6] In practice, however, the Committee failed to focus attention on the bases plan when it was referred to it in March. The subsequent entry of the Soviet Union into the war on Britain's side in June produced a further delay. It was felt inappropriate to consider postwar issues behind the back of an ally.[7]

Subsequent attempts were made to consider the bases plan by the Future (Operational) Planning Staff (a subdivision of the Joint Planning Staff) in early 1942 and by the Ministerial Sub-Committee (of the Committee on Reconstruction Problems) in the summer of 1942.[8] In both cases, however, despite Foreign Office interest in the idea, the military planners failed to provide any strong backing. In November the War Cabinet postponed a decision on postwar bases and the idea was not revived again until mid-1944.[9]

For the Foreign Office the need to make some progress on broader postwar planning was highlighted by a visit by Richard Law, the Minister of State for Foreign Affairs and Nigel Ronald, Assistant Under-Secretary of State, to the United States in August 1942. As a result of their discussions with American officials on the concept of a United Nations organisation, four-power control of Germany and the postwar settlement in Europe, it became clear that Britain would need to formulate its own views on these subjects. This led to the setting up of the Economic and Reconstruction Department in the Foreign Office under Ronald to consider long-term policy.[10] Before long, key officials in the Department like Gladwyn Jebb,

Samuel Hood and Evelyn Baring were producing studies which attempted to focus attention more broadly on British postwar policy. According to Jebb, by the summer of 1942, Foreign Office opinion was not only beginning to be suspicious of the Soviet Union but was moving towards a consensus that some form of West European group would be necessary in the postwar period. It was increasingly felt that Britain should attempt to organise some measure of unity in Western Europe, if possible with the cooperation of France and the United States. It was recognised that this would leave Eastern Europe to be guided either by some association of West Slavs with Russia in the background, or possibly (if Britain could not prevent it) by Russia herself. The major threat, however, as far as the Foreign Office was concerned, remained that of a resurgent Germany; the Economic and Reconstruction Department spent a great deal of time debating the rearrangement of German frontiers and the dismemberment of Germany. In line with American thinking it was also thought desirable for some form of international organisation to be set up in the postwar period based on great-power cooperation.[11]

Between the autumn of 1942 and the spring of 1943 Foreign Office officials wrestled with these ideas and some of the dilemmas which they created. One major problem centred on the relationship between a Western European association and the maintenance of great power cooperation in a postwar international organisation. In the autumn of 1942 Gladwyn Jebb produced his 'Four Power Plan' which urged the need for collaboration and good will between Britain, the United States, the Soviet Union and China in the postwar period.[12] Jebb went on to argue that if cooperation with the Soviet Union proved impossible then 'Britain would be driven into forming some kind of anti-Soviet front, and in so doing we should have eventually to accept the collaboration of Germany'.[13] Although the 'Four Power Plan' was accepted by the War Cabinet, the Foreign Secretary Anthony Eden remained uneasy about the idea. In January 1943, he wrote a memorandum in which he warned of the possible inconsistencies between the aims contained in Jebb's paper.[14] Eden accepted the importance of Britain playing a role in Europe after the war to prevent the domination of the continent by any one power. However, he also argued that the establishment of regional blocs by one or other of the great powers would be likely to lead to rivalry and could be the cause of a future war. Regional groups, he maintained, could only be useful within a world security system and it was important that the great powers should continue to be equally interested in the maintenance of peace everywhere in the world.

Eden attempted to confront the dilemma over postwar Germany in March 1943 in a paper to the War Cabinet.[15] He argued that the basis of

British policy should be the disarmament of Germany and the prevention of German rearmament. He was concerned, nevertheless, to remind his colleagues that the experience of the First World War showed that care would have to be taken in dealing with Germany. In the early 1920s German opinion had been divided over whether the future of the country lay in collaboration with East or West. This would be likely, he warned, to happen again. It was important, therefore, that the British government should not lose sight of the possibility that, in order to forestall any orientation of German policy towards the USSR and the conclusion of a Russo-German alliance directed against the West, 'it might be necessary to convince the German people that it was in their interests in the long-term to associate with Western Europe'.[16]

Up to this point (the spring of 1943) Foreign Office planning for the postwar period therefore centred on attempts to prevent a resurgence of German power and to avoid any situation in which Germany might be used by the Soviet Union for her own purposes and so tip the balance in Europe to the disadvantage of Britain. It was generally hoped that this might be achieved through cooperation among the Big Three; by friendly relations between Britain and the states of Western Europe (especially France); and by the creation of confederations in central and Eastern Europe which, it was hoped, would not be dominated by the Soviet Union.[17] A rearmed Germany was still seen as the major threat, but suspicions of the Soviet Union were increasing. During his visit to the United States in May 1943 the Prime Minister, Winston Churchill, went out of his way to stress the need for a strong France 'since otherwise there would be no strong country between Great Britain and the USSR'.[18]

However, with the end of the war not yet in sight, Foreign Office planning still remained rather unsettled and inevitably imprecise in certain respects towards postwar Europe. What was meant by 'confederations' in Eastern Europe? The Secretary of State was putting forward suggestions for some form of 'UN Commission for Europe' and the Prime Minister had proposed a 'Council of Europe'. But what form would these bodies take? What kind of Western European group should Britain try to establish? The government's position on Germany had also remained extremely fluid. Should it pursue a policy of dismemberment? How could a united Germany be kept out of the Soviet orbit? Could Big Three cooperation be maintained in the postwar period? All these questions still remained unanswered.

As the Foreign Office planners continued to wrestle with postwar policy there was no shortage of advice from allies and governments-in-exile in London. One of the first public pronouncements of the need to draw the nations of Western Europe into closer association with the British Commonwealth came from General Smuts, the South African Prime Minister, in an address to the Empire Parliamentary Association on 25 November 1943.[19] Smuts recommended that the Western European nations should align themselves with Britain for their own good as well as for Britain's. Although he failed to define precisely the nature of the alignment he foresaw, he did pursue the idea a little further in a message to the British Prime Minister in March 1944 and in a conversation with General Montgomery in May of the same year. In his communication to Churchill he urged the Foreign Office to prepare a paper on the 'traditional trends of Russian policy' for the forthcoming Dominion Prime Ministers Conference.[20] In line with this he suggested that consideration should be given to the possibilities of establishing a regional grouping of Western European states around Britain. He made much the same point in his talk with Montgomery in May.[21] According to Montgomery, Smuts was worried that Britain and the United States might win the war and lose the peace. He urged Britain not to stand back from Europe as it had done after the 1914–18 war. Montgomery recalls in his memoirs that Smuts went on to say that Britain must not allow Europe to disintegrate. Europe, he suggested, required a structure on which to rebuild itself and 'a good structure must have a firm core'. France had failed dismally in his eyes and consequently Britain must stand forward 'as the cornerstone of the new structure'. There could be no neutrals, he argued. 'Nations that want security must range themselves on the side of Britain.'

The British government was getting similar advice at much the same time from various European leaders, many of them in 'governments-in-exile' based in London. A great deal of interest had been caused by the Prime Minister's 'United States of Europe' speech and not surprisingly many of the smaller Western European 'governments' in particular wanted to know more about Britain's future policies towards Europe. They also urged Britain to take the lead in making proposals for greater cooperation between Britain and the continent. In October 1942 Trygve Lie and Van Kleffens, the Dutch Foreign Minister, raised the question of Britain's attitude to postwar European security directly with the British government.[22] Lie once again argued that it was essential for the security of Norway and the future peace of Europe that permanent measures should be taken for the defence of the Atlantic. Van Kleffens (who was destined to play an important role in the negotiations which established the North Atlantic Treaty Organisation in 1949) informed the British government that Belgium, Norway and the Netherlands were considering whether to submit joint proposals to Britain and the United States on much the same subject. The Dutch Foreign Minister had also been in touch with the American Secretary of State, Cordell Hull, and his deputy, Sumner Welles, on the need for similar regional defence arrangements for the Pacific and Indian Oceans, the Mediterranean, and the Baltic, as well as the Atlantic. These proposals, had been favourably received by the United States.

Despite the continuing interest of the smaller European states in postwar European security, little was done to follow up such initiatives during 1943. In a conversation with Anthony Eden on 23 March 1944, however, Paul-Henri Spaak, the Belgian Foreign Minister, informed his British counterpart that the Belgian and Netherlands governments were determined to collaborate more closely on financial and military matters and they hoped that other Western European states, including Great Britain, would participate.[23] Spaak argued in favour of a Western bloc which would include Britain and would extend from Norway to the Iberian peninsula. He told Eden that the Belgian government believed that Britain did not realise how much the smaller powers looked to them to take the initiative and he urged the Foreign Secretary to 'speak out more clearly' on Britain's policy towards Western Europe.

Eden's reply revealed the continuing uncertainty and caution in government circles on British plans for postwar Western Europe. The Foreign Secretary pointed out to Spaak that such proposals raised the difficult problem of spheres of influence in Europe which might seriously undermine allied unity both in the war and its aftermath, with all of the implications that would arise from this. He did, however, agree to consider Spaak's views and reply in due course.

This conversation was followed up on 24 March by a memorandum from Spaak to Eden.[24] In it, Spaak accepted the dangers of creating two blocs in Europe but pointed out that the Soviet Union already seemed to be engaged in creating its own sphere in Eastern Europe. It was therefore important for Britain to react in some way to this situation. He urged the British government to make it clear to the Russians that peace could only be achieved by the organisation of Europe as a whole. They might suggest to the Soviet leaders that the Anglo-Soviet alliance should form the basis for a larger European system after the war. In the meantime Spaak urged the British government to play its part in any efforts the Belgian, French, Netherlands and Luxembourg governments might make to set up a Western economic and military organisation.

At much the same time, the Foreign Secretary also received a forceful recommendation in favour of a Western European security grouping from Duff Cooper, at the time British representative to the French Committee of Liberation in Algiers. Cooper, who had unavailingly urged the need for a continental commitment in the mid-1930s, produced a lengthy memorandum on 30 May 1944, in which he reviewed the world situation that he thought Britain would face after the war and advocated the kind of foreign policy he would like Britain to adopt.[25]

While supporting both the creation of an international organisation and continuing close friendship with the United States, Duff Cooper argued that Britain could not stake her future on either. It was to Europe that Britain must look. It was to be assumed that the mistake of allowing Germany to rearm would not be repeated, in which case he foresaw that Russia would pose the 'gravest potential menace to the peace of the continent' (though he did not rule out the possibility of an alliance between the two states). In these circumstances, he pointed out, France's search for security might lead her to look to Russia rather than to Great Britain, and smaller Western powers would be inclined to follow the lead of France. He argued, therefore, that Britain should act quickly to reassure the French and lay the foundations of a Western European system. Once again, like others, he urged that the first step must be the conclusion of a treaty of alliance with France to form the cornerstone of the vast edifice.

Duff Cooper's dream was a grand one. He saw the alliance between Britain and France gradually leading to a 'federation of the Western seaboard of Europe together with the principal powers of the Mediterranean'. He argued that of the three great world power combines, this federation would become the strongest – 'an alliance so mighty that no power on earth would have dared to challenge it'.[26]

Eden was interested in the ideas expressed in the memorandum but his reply suggested to Cooper that his views in general did not 'recommend themselves to the powers that be'. Eden pointed out that such an alliance of Western democracies would have serious effects on the main tenets of British policy, which was based on the need for continued collaboration with the United States and the USSR. It would be likely, he argued, to increase the danger, 'if indeed such a danger existed', of the Soviet Union pursuing a policy of expansionism in Europe. He also suggested that such a policy might well offend the United States.[27]

Cooper records in his memoirs that this was 'the old familiar attitude' – rather than risk offending anybody, do nothing.[28] In fact, as we have seen, the Foreign Office was already giving some thought to these ideas. Despite his criticisms, the Foreign Secretary himself reacted more positively to the memorandum than Duff Cooper realised. On 2 July Eden wrote a minute on Britain's relations with her neighbours, suggesting that the time had perhaps come for some action. In a rhetorical question the Foreign Secretary asked, 'what are we doing to discuss these matters with our Western associates? So far as I know, nothing at all. I have never spoken on these matters to Van Kleffens (Netherlands Foreign Minister), Lie, Spaak or

Massigli (French Commissioner for Foreign Affairs)'.[29] [Eden seems to have forgotten his discussion with M. Spaak on 23 March 1944.] He went on:

> They have not been encouraged in their various timid advances and they may soon return to their lands. I trust that they will, but then shall we not have missed an exceptional opportunity? Should I not speak to them on these things, and, openly, and soon? Should I not later invite Massigili and associate him with our discussions? Should I not tell USA and USSR what I propose? It is no doubt good that many papers should be prepared, but this seems an occasion for actions, and I should like a meeting to discuss it.[30]

By the summer of 1944 Eden seems thus to have become convinced of the need to discuss mutual defence arrangements as soon as possible. It was no excuse, he argued, for Britain to be without a foreign policy on Europe simply because she was waiting for the United States. Some kind of defence arrangements, he believed, were 'indispensable, whatever form the proposed world organization took'.[31]

In a letter to Duff Cooper on 25 July he argued that any world organisation which might be constituted would have to be reinforced by various systems of alliances. He remained, however, sensitive to the problem of not alienating the Soviet Union. He had been aware for some time of the danger of the Soviet Union pursuing an expansionist policy in Europe but he believed more than ever that it was 'important that any proposal for closer association between ourselves and the Western European allies, or even with the states of Western Europe, should be for the sole purposes of preventing a renewal of German aggression'.[32] It would be fatal, he believed, 'to let it be understood that there is any other purpose in such an association.'[33] He therefore considered that any durable system in Western Europe must be based on three things: the Anglo-Soviet alliance; the public commitment never to allow the revival of a powerful Germany; and an alliance system *within* a world organisation or close understanding between the Big Three.

The Foreign Secretary's views at this time to a large extent reflected the work that was going on in the Foreign Office to provide a brief for the British delegation at the talks scheduled to be held at Dumbarton Oaks in August 1944 on the framework for a world organisation. As part of this preparation two papers produced by Jebb in May on the question of a Western bloc were combined into a single memorandum (entitled 'Western Europe') which was sent both to the Secretary of State on 20 June 1944 and, with his approval, to the Chiefs of Staff for their views.[34]

The 'combined memorandum' pointed out that it was likely that regional security issues would be discussed in Washington. As a result it was important that consideration should be given to the implications of British policy towards Western Europe for relations with the US, the Soviet Union and the Dominions as well as for our own security interests.

As far as the United States was concerned it was widely recognised in the Foreign Office that American policy towards a Western European security scheme was somewhat ambiguous.[35] There were those in the United States who were very suspicious of any attempt to establish a closer grouping of Western European states before the creation of a world organisation. There were also those in the administration who were very critical of proposals which tended to divide the world into blocs or spheres of influence. And there were isolationists who argued that if the Europeans organised themselves in such a way there would be no need for American involvement in postwar Europe. At the same time there were others in the United States who regarded Britain and Western Europe as an important area for future American security, as an 'advanced outpost', and who were more likely to welcome moves which supplemented British power and security.

In discussing these differing American attitudes towards a Western European group, the new memorandum emphasised the vital importance to Britain of maintaining close Anglo-American relations in the postwar period. It was therefore important to carry the Americans along with any proposals for a Western European security system and it was felt that this could best be done through emphasising the United Nations framework. It was also suggested that every effort be made to persuade them to accept a European commitment after the war ended. It was thought that the latter might not be possible, but while the chance remained everything should be done to try and achieve it.[36]

The memorandum also dealt in some detail with the need for careful thinking about the implications for relations with the Soviet Union, on which the Foreign Secretary had put so much stress. As with the potential American reaction, it was difficult to know how the Soviet government would react to British attempts to organise Western Europe. Some of the evidence suggested they might willingly accept such proposals. Stalin had, after all, suggested in late 1941 that Britain should assume certain defence obligations in Western Europe. Since then, however, difficulties had arisen in Anglo-Soviet relations at various times and it was recognised that any attempts to work outside the framework of a world organisation on an exclusively Western European security system would probably be seen by Soviet leaders as a move directed against them. The memorandum therefore concluded (once again) that every effort would have to be made to set Western European plans in a wider framework and to emphasise that the objective of any regional groupings would be to prevent the reassertion of German power.[37]

It was recognised that sensitivity would also be required in the way the plan was broached with the Dominion leaders. The Dominion prime ministers had already shown in their May 1944 Conference in London that they were not happy about Britain binding herself too closely to Europe by assuming much greater continental commitments. They had expressed their concern about British policy involving them once more in a European war. At the same time, however, they had recognised the need for some European grouping within a world organisation and that precautions had to be taken against a new world war. They had also recognised the strategic reasons which dictated a closer association in future between Great Britain and the Western European states. In reflecting these views the memorandum also noted that the Dominion leaders were concerned that any system which was created should include the United States (towards whom they were increasingly looking for their own security), and if possible the Soviet Union as well.[38]

After discussing the delicate nature of these relations the 'combined memorandum' then went on to discuss the shape a regional security system might take and the advantages and disadvantages of such a system for Britain.

In terms of the postwar debate about how a Western European security system ought to be developed it is significant that as early as this (June 1944) two possible alternative arrangements were being considered. It was argued in the 'combined memorandum' that such an arrangement might take the form of a multilateral treaty of mutual defence which would include Britain, France, the Low Countries, Denmark, Norway and Iceland (and perhaps even Sweden, Spain, Portugal, and Italy). Alternatively, it might involve a series of bilateral treaties of mutual defence between Britain and these states and between the other states themselves.[39] If this proved difficult it was suggested that there might be joint staff talks and attempts to coordinate the defence plans of the various states. The memorandum was adamant, however, that any plan would have to be in line with, and subordinate to, the provisions of any world organisation system which was set up.

The memorandum then reiterated the arguments for and against a Western European regional security system which had appeared in the previous papers drawn up by Jebb. As far as the advantages were concerned, six main points were stressed. These included contributing to Britain's role within a world organisation; reinforcing American and Soviet expressed desires for Britain to assume defence obligations in Western Europe; reducing Britain's defence burden in the postwar period; helping to strengthen France; complementing the Anglo-Soviet treaty; and the enhancement of British security through 'defence in depth'.

Apart from some of the difficulties, already mentioned, which might arise in relations with the US, USSR and Dominions, the main disadvantage foreseen was the 'increased risk of getting involved in the defence of Western Europe and the maintenance of land forces on a continental scale'. Interwar thinking still prevailed. There was also the warning that British participation should not be at the expense of her overseas commitments.[40] The memorandum emphasised that it was important for Britain to retain the traditional balance between policy in Europe and the rest of the world and to keep in the closest association with the United States. These arguments, like those on the benefits, were to be heard time and time again in the postwar debates.

Overall the memorandum concluded that Britain's political and strategic interests did require some drawing-together of the states of Western Europe and that Britain should be part of any group that was formed. This conclusion was reached because it was felt that 'a weak and divided Europe would be a potential storm centre which might be exploited by Russia'.[41] Despite the sensitivity over American, Soviet and Dominion opinion, therefore, by June 1944 there was a strong body of opinion in the Foreign Office in favour of a firm commitment by Britain to Western Europe following the end of the war. Although there was a growing anxiety that 'a weak and divided Europe would be a potential storm centre which might be exploited by Russia', the Foreign Office remained primarily concerned with a resurgence of German power. Great-power cooperation would be an important requirement in the postwar world and any Western European grouping would have to be set in the context of the United Nations framework. This was not, however, a perspective which was wholly acceptable to the military planners at the time.

2 Emerging Differences between the Chiefs of Staff and the Foreign Office, 1944–45

Although generally supportive of a Western European regional security system, the position adopted by the Chiefs of Staff towards Germany, the Soviet Union and Foreign Office proposals for a United Nations organisation had been the source of some friction with the Foreign Office for some time. Following the failure of previous arrangements to achieve effective coordination of diplomatic and military planning for the postwar period, the Post-Hostilities Planning Sub-Committee (of the CoS) was set up in July 1943 with Gladwyn Jebb as its chairman.[1] Although its purpose was to provide a better framework for the coordination of military and diplomatic planning for postwar security, in practice this did not happen. Although the Post-Hostilities Planning Sub-Committee produced over sixty studies, increasingly the frictions which had been evident since 1942 between the Foreign Office and the CoS reemerged. Jebb's role as chairman of the Sub-Committee was regarded increasingly as 'diplomatic infiltration' by the Service Chiefs.[2]

A major dispute broke out in early 1944 when the CoS provided their verdict on the Four Power Plan. On 17 February the CoS met to consider the PHP paper on 'The Military Aspect of any Post-war Security Organisation'.[3] They were extremely critical of the arrangements for a World Council and the Military Staff Committee proposal contained in the paper which they regarded as 'quite unworkable'. Jebb, who attended the meeting, records that 'it was pretty clear that the Chiefs of Staff for their part did *not* accept the Four Power thesis'.[4] They argued that what in practice was likely to happen was that the Combined Chiefs of Staff would continue in being; that the Russians would have a very large sphere in which they would have their own 'security' organisation; and that China was 'anyhow rather a joke'. The CoS response was to send the paper back to the sub-committee with the instruction that it should be rewritten.

Eden, on the other hand, thought the paper was an 'excellent effort' and the position adopted by the CoS caused considerable irritation in the Foreign Office.[5] A head-on clash was only averted by a decision to reform the PHP and to set up a modified Post-Hostilities Planning Staff (PHPS) which would be responsible to the Vice-Chiefs of Staff rather than the CoS themselves.

Despite this further reorganisation in the machinery for postwar planning, over the summer and autumn of 1944 the disputes gradually became more pronounced. On 7 July the PHPS produced a draft report on British strategic interests in the North Atlantic and Western Europe.[6] The report was based on two basic assumptions; firstly that Three Power cooperation under a world organisation would continue; and secondly that tripartite discord would occur and cooperation would break down. On the first assumption it was argued that there would be no threat to British strategic interests in the area and the only problem would be to enforce Peace Treaty provisions on a disarmed Germany after occupation forces had been withdrawn. In line with the ideas put forward in 1942, it was suggested that 'air action would be the most potent method of deterrence and retribution – economical, speedy and psychologically effective'.[7] If, however, this proved politically difficult because of the opposition of public opinion it would be necessary to reoccupy Germany using land forces. An eastern group of states led by the Soviet Union and a Western group of states led by Great Britain and backed by the United States and Canada would have to be mobilised for this task.

Britain's contribution to the Western European group, it was argued, would be limited because of her world-wide responsibilities but token British forces on the continent would be necessary to demonstrate Britain's political determination to resist a resurgence of Germany. According to the draft, 'from our point of view, it would be a fatal mistake to refuse such undertakings if they are the only means which will secure the wholehearted cooperation of France, Belgium, Holland, Denmark and Norway and thus ensure the success of preventive action against Germany'.[8] At the same time it was argued that some form of American contribution based in the United Kingdom or on the continent would be an important signal of United States interest in the security of Western Europe. Such direct US involvement, however, would not be essential if Three Power harmony persisted.

The second assumption, it was contended, presented rather more difficult problems for Britain. Tripartite cooperation could break down for a variety of reasons, and if it did, a Western European defence system would become even more important. Britain would have to make a military contribution to continental defence which, however small, 'would be out of all proportion to its size and could well prove decisive in ensuring Western European cooperation'.[9] The possibility that the United States might fail to intervene meant that Britain would have to be ready to attack Germany 'by air, sea and land on the first sign of rearmament'.[10]

The possibility that the Soviet Union might be hostile was treated in a rather cursory way. A Russian advance across Europe could not be halted and Britain would have to fall back to the home islands as had happened at Dunkirk. A long war would follow and only the United States would be capable of restoring the situation. An even worse threat would arise as a result of a Soviet–German alliance and this had to be avoided 'at all costs'. In certain circumstances, however unpalatable, Germany might have to be brought back into the Western European group to balance the power of the Soviet Union.

Although the draft was generally well-received in the Foreign Office because of the commitment to a Western defence group under either assumption, there was some criticism from the Northern Department of the attitude adopted towards the Soviet Union. Warner, the head of the Department, complained that the paper barely concealed the thought that Russia was just as likely to be an enemy as Germany and that it might be necessary to combine with Germany against the Soviet Union. He emphasised the argument that the Foreign Office view was that 'the most important point in securing Russian collaboration after the war will be to convince Russia of our determination to go with her in holding Germany down and that only in the event of our appearing to play with Germany against Russia is Russia likely to try to get in first with Germany'.[11]

Despite Warner's criticisms, however, the paper was endorsed by the Foreign Office and passed to the Chiefs of Staff for their consideration. When they looked at it on 26 July their reaction was extremely critical. According to one PHPS secretary 'they shot it down, if not in flames at any rate with a smell of burning'.[12] The CoS argued the paper failed 'to face up to the hard military facts of the problem'.[13] They accepted that efforts had to be made to try and make the proposed world organisation a success and that preventing the resurgence of Germany was the primary objective. Much more attention, however, had to be given to 'the real military problem' which would arise from the collapse of the world organisation.[14] The CoS contended that 'we must on no account antagonise Russia by giving the appearance of building up the Western European block against her . . . for this reason the immediate object of a Western European group must be the keeping down of Germany; but we feel that the more remote, but more dangerous, possibility of a hostile Russia making use of the resources of Germany must not be lost sight of, and that any measures which we now take should be tested by whether or not they help to prevent that contingency ever arising'.[15]

The CoS believed that after the defeat of Germany the Soviet Union would be the largest land power on the continent and therefore had to be regarded as the principal potential enemy of the United Kingdom. Rather than weakening the section on the Soviet Union, as Warner had advocated, they wanted the PHPS paper strengthened to reflect 'both the worst and the most likely contingency'.[16] As a result the CoS sent the paper back to be rewritten.

The direction of the thinking of the CoS at this time is revealed in a diary entry on 27 July made by General Sir Alan Brooke, the Chief of the Imperial General Staff. He argued that Germany now had to be thought of in a very different light from that of the past few years. Germany was no longer the dominating power in Europe. The Soviet Union, with its vast resources, could not fail to become the main threat in the next fifteen years. The task, he said, was to 'foster Germany, gradually build her up and bring her into a Federation of Western Europe. Unfortunately, this must all be done under the cloak of a holy alliance between England, Russia and America. Not an easy policy, and one requiring a super Foreign Secretary.'[17]

The Foreign Secretary involved, however, was very much opposed to this line of argument. In a letter to Duff Cooper on 25 July he argued that the idea of a Western group organised as a defensive measure against the possibility of Russia embarking at some future date on a policy of aggression and domination would be a 'dangerous experiment'.[18] Echoing Warner's criticism of the PHPS paper, he warned that a policy of this kind would be likely to 'precipitate the evils against which it was intended to guard'.[19]

This was a view widely shared in the Foreign Office at this time. Although Jebb took some comfort from the fact that both the Foreign Office and the CoS agreed on the need for a Western European defensive group, he considered the policy of building up Britain's enemies to defeat her allies as deriving from some kind of 'suicidal mania'.[20] Frank Roberts of the Central Department argued that the CoS 'were not only crossing their bridge before they come to it but are even constructing the bridge in order to cross it'.[21] There were also comments from other officials about the 'cursory and dangerous' views of the CoS, their 'anti-Russian extravagances' and their 'almost fascist assumptions'.[22]

Following the problems over the July PHPS paper the dispute reemerged in September and October over the question of whether the dismemberment of Germany in the postwar period would be in the best interests of Britain. In a memorandum to the Foreign Office on 9 September 1944, the CoS supported dismemberment because, they argued, it would help to prevent

German rearmament and renewed aggression.[23] They also regarded it as 'an insurance against the possibility of a hostile USSR'.[24] Once again they argued that Britain had 'to guard against a Russo-German combination, and, in the event of Russian hostility, we should need German help'.[25] According to the CoS, the Russian leaders would be unlikely to allow the rearmament of a united Germany unless they believed they could dominate it. It was thus in Britain's interests to seek dismemberment 'since we might be able to bring at least north-west, and possibly also south Germany within the orbit of a Western European group'.[26]

By this time, however, the Foreign Office was becoming increasingly hostile to the notion of the dismemberment of Germany. In a memorandum to the 'Armistice and Post-War Committee' it argued that Germany would be likely to try to get round such a policy and British, as well as American, opinion would come to regard it as an injustice.[27]

The Foreign Office was singularly unimpressed by the arguments put forward by the CoS. Despite similar sentiments which had circulated earlier in the Foreign Office, it regarded the CoS suggestion that Britain might revise its policy towards Germany and use part of that country against a combination of Russia and the East Germans as 'fantastic and dangerous'. The Foreign Office now emphasised that British policy must be to maintain the unity of the great powers. In a memorandum written on 20 September Eden argued that any attempt to formulate plans with the idea that Germany might serve as part of a postwar anti-Soviet bloc, would destroy any chances of preserving the Anglo-Soviet alliance. Britain would soon find itself 'relaxing disarmament and other measures which we regarded as necessary to prevent future German aggression'.[28]

Once again the Chiefs of Staff replied to the Foreign Office that they realised the importance of trying to maintain friendship with the Russians and they hoped the world security organisation would succeed.[29] However, they had to take into account the possibility that relations with the Russians would not continue on friendly lines and that the hopes for peace through world organisation might not materialise.

In their reply the Foreign Office agreed that the possibility of war with the Russians could not be ruled out but they argued that such conflict was not likely for some time. It was therefore important to pursue friendly relations with the Russians and that this could be done without harming British interests. Such a relationship would not be possible if the Russian leaders came to the conclusion that Britain was trying to build up a bloc against them which would include Germany or any part of that country.[30]

In an attempt to resolve this difference of opinion, the Foreign Secretary himself met the Chiefs of Staff on 4 October for a final 'shown-down on the issue'. As before, the Chiefs of Staff argued that they recognised the value of friendship with the Soviet Union and a successful world organisation. However, it was their duty as Chiefs of Staff 'to examine all serious eventualities. We cannot be debarred from taking into account the possibility that for some reason or other the world security organization may break down, and that Russia may start forth on a path to world domination, as other continental nations have done before her.'[31] In their view a failure to consider this would lead to a repetition of the disastrous error of the interwar years when all planning had been based on the pious hope that the League of Nations would prevent aggression.

The Chiefs argued that the examination of an unpleasant situation which might arise in the future was not incompatible with the pursuit of a policy designed to prevent that situation from arising. They expressed their concern that the Foreign Office seemed to recoil from the precaution of considering how to insure against the failures of policy.

Eden in turn repeated the Foreign Office line that it was vitally important not to be seen to be ganging up on the Soviet Union.[32] If the Soviet Union discovered the existence of such planning it would precipitate moves which would render Western bloc aspirations ineffective. It would also make the Soviet Union determined to ensure that Germany was on their side rather than Britain's. The Foreign Secretary argued that the Soviet Union was in an excellent position to achieve both aims but it was unlikely to try 'unless they were provoked by the Service Departments'.[33]

The meeting between Eden and the CoS did little to resolve the divergence of views.[34] The CoS merely agreed to a restricted circulation of papers which referred to the problems with Russia. Ironically by the time the new code on circulating papers came into force on 14 November many of the papers on postwar planning had already been passed to the British Joint Staff mission in Washington via Donald Maclean, the Soviet spy who worked at the British embassy.

The 'fundamental cleavage' between the Foreign Office and the CoS reappeared once again in November with the completion of the revised PHPS paper on 'Security in Western Europe and the North Atlantic'.[35] The CoS had been 'fully roused' by the first paper on this subject in July – which they believed had not addressed the real military problem.[36] The reason for this they suspected was the Foreign Office 'infiltration' of the PHPS through Jebb's chairmanship. As a result, over the summer, Jebb was replaced as chairman and became merely an associate member of the the PHPS. When the revised version of the paper appeared on 9 November, some concessions had been made to the Foreign Office over the question of German assistance. The reduced influence of the

Foreign Office in the PHPS, however, was revealed in the increased emphasis now given to military planning to meet possible Soviet hostility.[37]

In a section on 'General Strategic Considerations' the report attempted to emphasise Britain's changed strategic position, likely future threats and security requirements in the postwar world. The development of air-power and long-range missiles meant that Britain's strategic position had seriously deteriorated. It was inevitable, the report said, that Britain had 'already become and will remain from the strategic point of view, much more closely bound to the Continent than hitherto'. The possible threats to Britain were seen to come from a rearmed Germany, the USSR or 'worst of all', both. In the light of Britain's traditional policy of maintaining a balance of power with the object of ensuring that no single great power was capable of dominating the continent, the problem in future would be that no single power in Europe would be able to balance the USSR, which had emerged as 'the greatest land power in the world'. As a result, the report stressed, Britain would need 'as never before', powerful allies; depth in defence, both on the continent and within the UK; and technical superiority in the application of science to warfare.[38]

The Post-Hostilities Planning Staff then considered Britain's future security problems on the basis of the same two alternative assumptions looked at in the first draft in July: that a world organisation would be set up and would prove to be effective and that such an organisation would either fail to materialise or, having materialised, would fail to be effective.

On the first assumption, it was argued again that an effective world organisation would inevitably involve close association between the United Kingdom, the United States and the USSR. If this were so the serious military problem envisaged in Western Europe and the Atlantic would be the prevention of German rearmament. If such a threat developed again, prompt and effective action, if necessary of a military nature, would be required. This would necessitate active British cooperation with France in particular, and with Belgium, Holland, Denmark and, possibly, Sweden. There would be several advantages in such cooperation. Among the most important would be that if reoccupation of Germany became necessary it would be against Britain's interests to allow the USSR to play a predominant role. Only through active cooperation, however, could the Western European states, including Britain, prevent this. But if Britain wanted to achieve the cooperation of the Western European states it would be necessary to demonstrate to them her determination and ability to take a full share in the task of keeping the peace, especially in terms of accepting a commitment to employ her armed forces in cooperation with allies against Germany.

The uncertainty prevailing in certain quarters in Britain, especially among the CoS, over whether the proposed world organisation would be set up, and if it was, whether it would prove effective, was reflected in the second assumption discussed in the paper, which might occur either because the US refused to undertake any definite commitments (or, having undertaken them, subsequently repudiated them) or through non-cooperation by the USSR. In a prescient section, the paper went on to point out that 'whatever the cause of a breakdown it is likely that there will be a reluctance to admit failure of an organization upon which so many hopes had been based'. This would mean that in the period between the beginning of allied disintegration and the final recognition of the fact by world opinion, Britain's security would be threatened. This being the case, the paper maintained that it was essential to pursue a policy which, while aiming to produce the necessary conditions in which a world organisation could come into being and be effective, would nevertheless provide Britain with security *just in case* the world organisation failed to materialise or broke down.

In such circumstances, it was recognised, threats to British security from a rearmed Germany or a hostile USSR would be much more difficult to deal with. In the case of a rearmed Germany, it might just be possible for the United Kingdom, in conjunction with the Commonwealth and in cooperation with the Western European states, to take the necessary preventive action. Combating a hostile USSR, however, would be significantly more difficult. Once again the need for cooperation with the Commonwealth and with the Western European states was fully recognised. In a conflict with the USSR 'defence in depth' would be essential. In the words of the report, 'the security of the United Kingdom could only be fully secured if we could deny to the enemy, and ourselves control, a wide belt of territory in Western Europe'.[39] Consequently it was emphasised that some form of Western European group including France, Belgium, Holland, Denmark and Norway as well as Sweden, Spain, Portugal and Iceland would be of great importance.

The paper went on to discuss the possibilities of securing some form of German cooperation (despite the dangers) should a Soviet threat materialise. Germany's geographic position and manpower resources were seen to be of crucial importance in such an event.[40] In line with contemporary Foreign Office thinking, however, it was emphasised that no attempt should be made to use Germany or part of that country against the USSR until relations with the latter had irrevocably broken down.

Even with an alliance including Western Germany, however, it was conceded that the forces available to stop the USSR would be totally inadequate. Echoing the views of the CoS, in particular, the PHPS, in a significant passage in the report, spelled out the necessity for close cooperation between any Western European Group which might be formed and the United States. The United States was seen as the only country possessing sufficient manpower and reserves to stabilise and restore a situation following a Soviet invasion of Western Europe.

Although the Chiefs of Staff refused to give unqualified approval to the paper, its sentiments were generally in line with their thinking. The Foreign Office, however, remained far from happy with the November PHPS paper. Gladwyn Jebb summed up the views of the Foreign Office when he commented on 21 November that 'the only result will be that the paper will be pigeon-holed and hardly any-one will see it. No doubt all to the good.'[41]

The fate of the two PHPS surveys on the North Atlantic area was similar to that of a number of the other regional studies they produced. Of the seven regional strategic papers they wrote only two (one on the Mediterranean and one on the Far East) were formally endorsed by the Chiefs of Staff or the Vice-Chiefs of Staff. Most of the others were either rejected outright or merely noted by the service chiefs. Part of the reason for this was that from the end of 1944, the regional studies were overshadowed by the more ambitious task of the PHPS of producing a world survey of Britain's postwar strategic requirements. The aim was to draw together the main points of the regional surveys to produce an overall grand strategy for Britain for the period from 1955 to 1960.

As part of the Foreign Office contribution to this global survey Jebb was asked to provide the political background against which military commitments could be planned. In line with the past disputes Jebb saw this as an opportunity 'to scotch the "Russian bogey" thesis' which prevailed amongst the service chiefs.[42] In his political forecast he predicted a 'troublesome and difficult' postwar period in Europe. Continuing civil disorder could give rise to the spread of communism but equally such unsettled circumstances, he argued, could lead to the emergence of populist dictators. It would be the responsibility of the three Great Powers to maintain order in Europe but the United States could not be relied on to keep its forces in Europe. In such circumstances, although the Soviet Union was unlikely to attempt to extend her control beyond a certain line, the countries of Eastern Europe would nevertheless probably be incorporated into the Soviet sphere of influence. Jebb accepted that Soviet probes in the Middle East and the Far East would have to be resisted. 'As for Anglo-American relations, these would remain completely interlocked, with the United Kingdom unable to contemplate any serious war without wholehearted American support and with the United States unable to abandon the British for fear of a situation developing in which American interests would be menaced by a hostile united Europe'.[43]

The Foreign Office wanted it clearly understood that despite their help in drafting the strategic survey they were 'in no way responsible for its views'. They accepted, however, that the paper might become the 'bible' for future service planning and therefore they had to try to influence the political context of such an important survey. In practice the Post-Hostilities planners eventually decided to omit the 'Political Forecast' completely because of the difficulty of prophesying the political situation in the 1955 to 1960 period. Instead it was decided to examine the potential danger of a hostile Russia alone while stressing the hope that 'there will never be any reason to believe that the USSR will become a hostile power'.[44] This led to pressure to insert a disclaimer which pointed out that the survey 'is written from a purely strategical point of view and that it is based upon hypotheses the probability of which is in no way endorsed by the Foreign Office'.[45] The authors, however, declined to insert such a paragraph arguing that the *possibility* of Soviet hostility had been the basis of all postwar planning for the past twelve months.

In line with previous PHPS papers, more than half the survey was therefore devoted to the potential threat from the Soviet Union. The rest dealt with the need to insure against Germany and Japan; minor threats and internal security difficulties; the United Nations organisations; and the question of how to secure greater cooperation with the Dominions.

After considering the possibility of clashes of interest with the Soviet Union in different parts of the world the global survey argued that the overall situation was potentially 'extremely grave':

> The USSR might commence hostilities with the limited objective of rapidly seizing some area of strategic important e.g. the Middle East oilfields, calculating that we should accept the situation rather than precipitate a world-wide war.
>
> In a full-scale conflict, however, the USSR could hope for the most decisive results in Western Europe, by attack on the United Kingdom and our vital Atlantic communications. The possibility of an airborne assault on Great Britain in the early stages of a war, will require reassessment from time to time. An attack on the Middle East, where our oil supplies and Mediterranean sea and air routes are within comparatively easy reach, would almost certainly be a feature of Soviet strategy. In India and the Indian Ocean, the USSR could contain large British forces without a great expenditure of effort.[46]

Faced with such a potential world-wide threat the survey argued that Britain needed a global strategy. In Europe, a Western group would need to be set up consisting of France, Norway, Denmark, Belgium, the Netherlands and Britain. Such a group would require the support of the United States and every effort would have to be made to prevent Germany from making common cause with the Soviet Union. Echoing the previous disputes with the Foreign Office the paper accepted that no attempt should be made to build up Germany against the Soviet Union until Anglo-Soviet relations had deteriorated beyond repair. In the Middle East mobile forces would be necessary to protect the wide range of British interests in the region. Air and civil defence systems would be needed to meet a possible airborne invasion in north-west India. In South East Asia and the Pacific efforts would have to be made to ensure the provision of air and naval bases and to deny offensive bases to the Russians.

The survey argued that if such a global strategic policy could be backed by the resources of all members of the British Empire then it could become a Great Power comparable with the United States and the Soviet Union. It was accepted, however, that so far the other Dominions were reluctant to undertake firm commitments and play their part in a coordinated imperial plan. The paper argued that this would have to be overcome by improved methods of imperial consultation. All members would benefit if they agreed 'to hold forces in readiness to serve in an imperial strategic reserve' and accept much 'greater coordination of military training, equipment and organisation'.[47]

In its conclusions the PHPS paper argued that the avoidance of a clash with the Soviet Union was of vital importance. It must be 'the primary object of policy'. It was also argued 'that the possession of military strength would further this aim; that imperial unity and co-ordination would be essential for the maintenance of Great Power Status; and that the full and early support in war of the United States forces would be essential for survival'.[48] In terms of the priorities between the various regions the defence of Western Europe and the security of Atlantic communications was deemed to come first. The second priority was accorded to the Indian Ocean area because of its vital role in imperial communications and as a source of manpower and industrial resources. The Middle East was regarded as the third priority because of the difficulty of defending the region against a Soviet assault. And finally came Far Eastern interests where the threat was relatively small and the United States could be expected to deal with Soviet probes.

When the PHPS paper finally went before the Chiefs of Staff on 12 July it received a rather mixed reception. The Chief of the Air Staff, Lord Portal, in particular questioned the decision to give such a higher priority to the Indian Ocean area rather than the Middle East. He also argued that greater attention should have been given to the need for alliances. General Brooke, on the other hand, thought that the survey was most useful and he thanked the Directors of the PHPS for 'the energy they had displayed in writing this most difficult paper and in handling the many and diverse questions which had been referred to them in the past'.[49]

To the relief of the Foreign Office, however, the Chiefs of Staff decided to merely 'take note' of the world survey rather than to formally approve it. The CoS accepted that the survey set out accurately the needs of imperial security but they regarded it as deficient as a guide for the detailed planning of the structure of postwar forces. Its main value, the Chiefs argued, would be as a strategic background against which the Joint Planning Staff could work out specific service requirements for the future.

Despite the continuing divergence between the Foreign Office and the CoS on how much emphasis to give to the Russian threat there was no fundamental disagreement between them on the need for a Western European Group in the postwar period. The Foreign Office continued to receive representations from the smaller Western European states throughout late 1944 and early 1945 in favour of such a group. These were received favourably but it proved impossible to take more positive steps in laying the foundations of a Western European Group because of a significant difference of opinion between the Foreign Office and the Prime Minister.[50]

Part of the problem stemmed from the disagreement which had arisen between Churchill and Eden over the form of a new world organisation. Churchill favoured the establishment of a world council to help maintain the association of the Great Powers (particularly Britain and the United States). However, such a council, he believed, would have a limited role. He placed great emphasis on regional councils which (despite their subordination to the world council) would have the main responsibility for keeping the peace in their areas. It was in this context that he emphasised the importance of a United States of Europe and a Council for Europe. The Foreign Office and Eden, on the other hand, supported a more universal organisation with an emphasis on 'smaller, specific-purpose regions'. From this point of view the World Council, rather than the various regional councils, should have 'final responsibility for the preservation of peace in every part of the world'. Foreign Office acceptance of regionalism, as we have seen, took the form of support for more limited groups such as a Western European security organisation which would be of advantage to

British interests in the event of the failure of the permanent world organisation.[51]

Churchill himself was not totally opposed to a Western European group but he was concerned to delay discussions for the time being. This was largely because he believed that the small states of Western Europe would probably be 'liabilities rather than assets if we bound ourselves to them in a scheme of common regional defence'. According to the Prime Minister in a minute to Eden of 25 November, there was little to be gained from such a group until France had been able to build up her army once again (which would probably take five to ten years).[52] The Prime Minister was particularly dismissive of many of the small states. 'The Belgians are extremely weak', he argued, 'and their behaviour before the war was shocking. The Dutch were entirely selfish and fought only when attacked, and then for a few hours. Denmark is helpless and defenceless, and Norway practically so. That England should undertake to defend these countries together with any help they may afford, before the French have the Second Army in Europe, seems to me contrary to all wisdom and even common prudence.'[53] His personal antagonism towards de Gaulle also militated against such a group. He was profoundly averse to any formal alliance with a France governed by de Gaulle.

Churchill went on to play down the strategic changes that had occurred in Britain's position and to deprecate the need for a strong continental presence in the postwar world. It could be that the continent might be able to fire at Britain and Britain at the continent. To that extent Britain's island position was damaged. A strong air force and sufficient naval power, the Prime Minister maintained, however, would continue to be a tremendous obstacle to invasion. As to the expense of keeping a large army on the continent, it would be wiser to put the bulk of the money into the air force 'which must be our chief defence with the Navy as an important assistant'. Clearly the Prime Minister himself was not yet convinced that the events of the 1930s and the war necessitated a fundamental change in British defence policy towards a greater continental commitment.[54]

In some important respects Churchill's position was closer to that of the Chiefs of Staff than the Foreign Office. In his view Britain's future lay in close relations with the United States, particularly in the military field. A Western European group might be of some value at a later stage but it should not be pursued at the expense of the Atlantic relationship. If a choice had to be made between the Foreign Office preference for British leadership of Western Europe and the CoS preference for close ties with the United States, Churchill undoubtedly favoured the latter.[55]

Absorbed as he was by the conduct of the war, Churchill was puzzled by all the talk of a Western bloc which was taking place at this time. He concluded his note to the Secretary of State by asking him how the various ideas on the subject had got around in the Foreign Office and other 'influential circles'.[56]

In his reply of 29 November Eden agreed with some of the Prime Minister's views on the danger of taking on widescale commitments on the continent.[57] He argued nevertheless that the time had come to start serious thinking about the problem of a regional defensive system. Echoing the sentiments of Duff Cooper's 'May memorandum' he suggested to the Prime Minister that it was important that the Western allies and France should not get the impression that Britain was not prepared to accept any continental commitments. If they did come to this conclusion they might be likely to seek accommodation and defence arrangements with the Soviet Union. The advent of long-range missiles, in Eden's view, really had altered Britain's strategic position and it was therefore necessary to achieve some form of 'defence in depth' to enhance Britain's future security. From this point of view a Western European security group could provide the necessary depth in defensive terms and the manpower to ease Britain's burden in the postwar world. It would not be necessary to maintain a large standing army on the continent, though a larger commitment than in the past would clearly be desirable. Eden then outlined for the Prime Minister the history of the discussions about a Western European security group which had been taking place for nearly two years in the Foreign Office and the views which had been expressed by the Chiefs of Staff on the subject.

Despite the Foreign Secretary's arguments in favour of a more positive attitude towards detailed discussions with the European states very little progress was in fact made in the later stages of the war or in the early months of the postwar period in Europe. There were a number of reasons for this. One of these centred on the continuing conviction of the Prime Minister that France, Belgium and the Netherlands were so desperately weak, and would remain so for some time, that any regional security system in which Britain would accept a commitment to defend them was not in Britain's own interests. Another major difficulty soon arose in negotiating an agreement with the French and General de Gaulle. Apart from the antagonism between Churchill and de Gaulle, there were also suspicions (raised by the Levant crisis and later problems of the frontiers with Germany) which proved to be stumbling blocks to negotiations of an Anglo-French treaty for some time to come.

Of central importance, however, in this lack of progress in planning for European security was the Prime Minister's preoccupation with the need to maintain close ties with the United States, particularly in the light of the growing difficulties with the Soviet Union. The American relationship presented a major problem for the British Prime Minister in the last two years of the war. It was one of the great disappointments of Churchill's career that he was unable to convince the Americans of the need to adopt a much tougher approach towards the Soviet Union during the final stages of the war. By the end of 1943 the Prime Minister had already concluded that Germany was finished, although it would 'take some time to clean up the mess'.[58] He told Harold Macmillan on his way to the Tehran Conference, 'The real problem now is Russia' but 'I can't get the Americans to see it'.[59] This arose largely from very different views on both sides of the Atlantic about the intentions of the Soviet Union and the prospects for continuing great-power collaboration in the postwar era. For the American President, Franklin D. Roosevelt, there was a need to avoid a return to the old balance-of-power politics and spheres of influence and establish a new era of world peace through a United Nations organisation. As part of this internationalist view everything had to be done to avoid the impression that Britain and the United States were 'ganging up on the Soviet Union'.[60] The United States and the Soviet Union would need to cooperate in the postwar period to maintain peace. At the same time colonialism was seen as a major impediment to the construction of the new world order. This implied criticism of the British Empire and suspicion that Britain intended to maintain its imperial possessions.

The British view, and more particularly Churchill's view, of the postwar order was rather different. To the extent that cooperation with the Soviet Union would be possible, Churchill believed that it had to be rooted in the realities of power. On 9 October the British Prime Minister met Stalin in an attempt to achieve an informal spheres-of-influence agreement in Eastern Europe. This involved a percentages deal in which Britain would have 90 per cent influence in Greece in exchange for 90 per cent Soviet influence in Romania. In Bulgaria it would be 75 per cent for the Soviet Union and 25 per cent for Britain. In Yugoslavia and Hungary each would have 50 per cent influence. For Churchill this represented an attempt to limit Soviet power and maintain British influence in key areas in Europe and before the changing balance of power on the continent weakened Britain's bargaining position.[61]

This attempt to do a deal with Stalin represented a growing realisation by Churchill that Roosevelt did not share his anxiety about Russian expansionism. This was reflected in growing disagreements over the conduct of the war and especially the Italian campaign. For Britain victory in Italy opened up the opportunity of playing a more influential role in central and southern

Europe and checking Soviet ambitions in the region. The United States, however, rejected British views, arguing that military considerations had to be paramount in the conduct of the campaign against Germany. The British were viewed in Washington as being overly concerned with carving out a political sphere of influence in Eastern Europe and the Balkans.

The divisions between Britain and the United States were evident at the Yalta Conference in February 1945. Despite Churchill's belief that 'only a solid understanding between the United States and Britain could keep Stalin's appetite under control',[62] Roosevelt refused to accept any prior coordination of Anglo-American policies. For Roosevelt the key objective was the establishment of the United Nations and he was determined to maintain close cooperation with the Soviet Union to achieve this objective. This approach tended to colour the deliberations between Stalin, Churchill and Roosevelt over Poland, which was the most important issue at Yalta. For Britain and the United States the future of Poland was of major importance. Britain had gone to war in 1939 when Poland was invaded by Hitler. The numerous Americans of Polish descent also encouraged Washington to give support to the Polish government-in-exile which had been set up in London. In July 1944, however, the Red Army had established a pro-Soviet government in Lublin and ruthlessly suppressed other nationalist forces which sought a more independent future for their country. At Yalta, although a rift was avoided, no final agreement on Poland was achieved. Britain failed to get American support on Eastern Europe and Churchill felt constrained in pushing the Polish case. By arguing that Britain would not tolerate any interference in the affairs of the British Empire the principle was established that the allies would not interfere in each other's spheres of interest. This left the Soviet Union a free hand in Eastern Europe by virtue of the control established in the region through the presence of the Red Army. The result was a far-from-satisfactory 'Declaration on Liberated Europe' which emphasised the importance of the principles of independence and sovereignty but which was vague about the means to achieve these ends. As far as Poland was concerned an agreement was reached whereby elements of the London government-in-exile would join the Lublin Communist government. Such an arrangement, however, only served to ensure Moscow's longer-term domination of Poland.[63]

Disagreements between Churchill and the Foreign Office continued after the Yalta Conference. Within two weeks of the 'Declaration on Liberated Europe' the Soviet Union engineered a *coup d'état* in Romania. The Russians also obstructed the work of the British mission in Bulgaria. Like the United States, the Foreign Office viewed these events as regrettable but they were not prepared to prejudice the wider objective of maintaining cooperation with the Soviet Union. Churchill, however, took a tougher line. He remained determined to oppose Russian actions and was unwilling to give up the battle to establish democratic governments in Eastern Europe and the Balkans.

Churchill expressed his anxieties about postwar cooperation with the Soviet Union to the War Cabinet on 3 April.

> It was by no means clear that we could count on Russia as a beneficent influence in Europe, or as a willing partner in maintaining the peace of the world. Yet, at the end of the war, Russia would be left in a position of preponderant power and influence throughout the whole of Europe. In the Western hemisphere, the United States had made enormous strides during the last two years, and had built up a military machine and supporting war production which was maintaining a vast military effort, not only in Europe but in the Pacific theatre. The resources in men and material commanded by the United States were vastly superior to our own; and they had acquired during this war a new capacity and experience in marshalling these resources in war.[64]

The Prime Minister believed that the advances being made by the Soviet Union in Eastern Europe could only be halted with the support of the United States.

Despite his efforts, however, the United States, even under a new President, Harry Truman, seemed oblivious to the dangers in Europe. Churchill's arguments for a rapid military advance on Berlin and the liberation of Prague to forestall occupation by the Red Army were rejected in Washington. The American government continued to oppose British plans on the grounds that they were motivated by political rather than military considerations.

Truman also appeared determined to continue Roosevelt's policy of not being seen to 'gang up' with Britain against the Soviet Union. Attempts by Eden to arrange a meeting with the new American Secretary of State, James Byrnes, prior to the Potsdam Conference in July 1945, to coordinate policies, were rebuffed. Although there was some convergence of views at the Conference itself, Britain and the United States remained divided on a number of key issues. The most serious, from the British point of view, concerned the attempts by the Soviet Union to secure the trusteeship of Tripolitania, an Italian colony in North Africa. Churchill saw this as a major threat to vital British interests in the Mediterranean and opposed Soviet demands. Despite Truman's willingness to take a tougher line with the Soviet Union over Eastern Europe, the Americans appeared unwilling to help protect Britain's interests in the Mediterranean. Suspicion of British imperialism in Washington continued to preclude a close partnership between the two countries.

By the time the Churchill government left office as a result of its defeat in the general election of July 1945 the Prime Minister's objective of securing American support to halt Soviet advances in Europe had largely failed. In Churchill's view it was already too late: 'the time to settle frontiers has gone. The Red Army is spreading over Europe. It will remain.'[65]

One of the main results of Churchill's preoccupation with the American relationship was that planning for a Western European security group became somewhat bogged down in the final stages of the war. As a result of the lack of clear political leadership at the top, the major bureaucratic battles of 1944 continued during 1945. Sensitivity over the attitudes of both the Americans and the Russians and concern not to undermine the prospects of the United Nations still remained important factors in Foreign Office thinking. Similarly, there was a continuing reluctance to open negotiations with the Western European states. Nevertheless, by 1945 from the Foreign Secretary down, there was a general acceptance in the Foreign Office of the advantages of a Western European security group and a recognition of the need for a more positive approach by Britain to the establishment of such a group. This view was generally accepted by the Chiefs of Staff but the emphasis in their planning, like the Prime Minister's, was rather more on the importance of continuing to maintain close military ties with the United States. For the CoS a global strategy would be needed in the postwar period to deal with the Soviet Union which was 'the only serious potential threat' to British interests all over the world. Western European security therefore had to be part of a much bigger grand strategy which could only contain Soviet power if the United States played its part in regional security arrangements alongside Britain. In contrast, by the end of the war, despite its growing anxieties about the Soviet Union, the Foreign Office remained rather more open-minded on the chances of Great Power cooperation succeeding. These issues of a global strategy, a special relationship with the United States and a West European security group, were to form the basis of the major debates in Britain about foreign and defence policy for the next four years of peace, and many more thereafter.[66]

3 Postwar Attitudes towards the Soviet Union

The change from a Conservative to a Labour government in July 1945 during the Potsdam Conference was less significant in the foreign policy field than it might have been because the new Prime Minister, Clement Attlee, and the new Foreign Secretary, Ernest Bevin, had played key roles in the wartime inner Cabinet. Both had been members of the Armistice and Post-War Committee which was the focus for discussions about the postwar period and which acted as a clearing house for proposals from the Foreign Office, the Chiefs of Staff and from a wide variety of official committees. It met 23 times between April and the end of 1944 and considered no fewer than 127 papers and reports. Alan Bullock has rightly argued that 'no better apprenticeship for a future Prime Minister and Foreign Secretary could have been devised'.[1] This continuity was reflected in the attitude towards the Soviet Union. In the months that followed the end of the war, growing difficulties with the Soviet Union over events in Eastern Europe and the Middle East exacerbated existing suspicions amongst Foreign Office officials. As those suspicions grew, Bevin was prepared to take a tough stand with the Soviet leaders. The new Foreign Secretary was a 'combative, tough-hewn, forceful' individual with a long experience in the Trades Union movement which had left a suspicion of Communism.[2] He went to Potsdam determined 'not to have Britain barged about'. Even more than his officials, however, he remained committed to his predecessor's objective of trying to maintain cooperation with the Soviet Union at least until the end of 1947.

In July 1945, on Eden's suggestion, Orme Sargent, Deputy Under Secretary of State, produced a memorandum on 'Stock-Taking after VE-Day'.[3] Britain, he argued, was in a much more difficult position than it had been after the Great War in 1918. It was essential to learn from the mistakes of the interwar period and come to terms with the realities of international life in 1945. The United States and the Soviet Union were now the dominant powers. Unless Britain asserted itself, neither of these powers would be prepared to consider British interests.

This could best be done, Sargent argued, by 'ceaselessly advocating the merits of three power co-operation, so as to make it difficult for the other two to ignore Britain'. Already there was a 'misconception' in Washington that Britain had been reduced to the ranks of a second-class power. The result of this was a tendency to disregard Britain and pursue an understanding with the Soviet Union. In order to take its rightful place alongside the other two world powers Britain would need to become the acknowledged leader of the Dominions and of the Western European states. 'Only so shall we be able, in the long run, to compel our two big partners to treat us as an equal.'[4]

Sargent acknowledged that continuing cooperation, especially with the Soviet Union, would be difficult. The objectives of Soviet policy were extremely difficult to discern. There was little to guide British policymakers in their assessment of Soviet aims. However, Stalin did appear to be particularly concerned about security and fearful of a resurgence of German power. It seemed likely therefore that he would try to secure those areas he believed to be vital for Soviet security either through territorial conquest or by the imposition of 'an ideological *Lebensraum*'. For the moment it could be assumed that he would not risk war, given the weakness of the Soviet Union. His approach would probably be to consolidate Soviet power in Eastern Europe but he could be challenged in areas like Poland, Czechoslovakia, Austria, Yugoslavia, Bulgaria and Finland. In Romania and Hungary, however, the Soviet Union was already in control and little could be done. Sargent argued that it would be important to challenge the Soviet Union where it was possible in Eastern Europe, not only for the sake of the countries concerned, but also to prevent Soviet pressure on Italy, Greece, Turkey and particularly Germany. The struggle for Germany would be of major importance. 'The result will be decisive for the whole of Europe, for it is not overstating the position to say that if Germany is won over to totalitarianism this may well decide the fate of Liberalism throughout the world.'[5]

Postwar difficulties lay not only with the Soviet Union, but Germany as well. There was a danger that Germany would 'recover its political independence, cast aside any democratic forms which its conquerors had imposed upon it and move to some form of authoritarian government, and, with equal rapidity, rebuild its mighty economy. If this happened Germany would then be in a position to 'put herself up to the highest bidder so as to play off each of the three Great Powers one against the other.' Britain would have no alternative but to play a greater role in Europe given these possible developments. It would also be necessary to convince the Americans of the futility of their present course of trying to mediate between Russia and Britain and 'to come into Europe wholeheartedly on the side of Britain'.[6]

Sargent's memorandum was received favourably within the Foreign Office and provided the basis for postwar policy towards the Soviet Union when Bevin took over as Foreign Secretary during the Potsdam Conference.

The emphasis was on the importance of continuing great-power cooperation, with a willingness on Britain's part to challenge the Soviet attempt to consolidate its power in Eastern Europe. This was very much the position that Bevin adopted in the months that followed.

At the first meeting of the Foreign Ministers of the Great Powers held in London in September 1945 the Foreign Secretary told Molotov that 'the chief difficulty' in relations between the two countries lay in Soviet policies in Bulgaria and Romania. He suggested that an independent inquiry should be set up to consider whether democratic liberties were being fully maintained in both countries.[7]

Molotov's response was to challenge Britain on the great-power status of France. Given the importance of France in Britain's wartime planning on Western Europe, Bevin continued his predecessors' support for complete equality for them in great-power deliberations about peace treaties with Germany and her allies. During the Conference, however, Molotov demanded that the discussion of the peace treaties should be confined to the Big Three, together with those countries actually at war with each individual enemy. This would mean that France would only be allowed to sit in when a treaty with Italy was being discussed. On other treaties France would be excluded. Bevin was adamantly opposed to this suggestion. Instead he suggested a compromise. Elections would be held in Romania and Bulgaria, 'on the same lines as Finland', where they had been free from Soviet interference.[8] France would also be accorded full equality in the drawing up of the satellite peace treaties. In return Britain would recognise the government which the Soviet Union had set up in Austria and Hungary. This was not, however, acceptable to the Soviet Union.

Disagreement also continued over Soviet demands for a ten-year trusteeship over Tripolitania, the western half of the former Italian colony of Libya, which was under British military occupation. Like Churchill, Bevin firmly rejected the demand. The Foreign Secretary and Chiefs of Staff were totally opposed to the idea of a Soviet colony in North Africa which they believed would threaten British interests in the Mediterranean and the Middle East. Every effort had to be made to prevent the Soviet Union from 'coming across the throat of the British commonwealth'.[9]

The deadlock at the London Conference was regarded with great dismay in the Foreign Office. The Russians were seen as being wholly opportunistic and self-seeking. Bevin's Private Secretary, Pierson Dixon, noted in his diary at the time that 'the depressing thing is the utterly realistic and selfish approach of the Russians and the complete absence of any wider consideration of the interests of peace for the best benefit of all'.[10] The failure of the Conference not surprisingly resulted in an increase in suspicion in the Foreign Office of Soviet aims in Europe and elsewhere. According to J. G. Ward, of the Economic and Reconstruction Department, the time had now passed when the Russians could be bought off 'by handing over to them the countries of Central Europe and the Balkans'. The Soviet demands for trusteeship in Tripolitania and for a share in the occupation of Japan were almost certainly designed to provide a springboard for 'a general campaign to establish Russian influence in every quarter of the globe'.[11]

Despite this dismay over the outcome of the London Conference Bevin remained determined to keep the differences private. After the Conference he gave instructions to avoid any public polemics against the Soviet Union. Unlike some of his advisers, he felt it was still too early to come to definite conclusions about the longer-term significance of Soviet behaviour. He believed it was important to allow 'time for things to simmer down and for the Soviet Union to show its hand more clearly'.[12]

While the Foreign Office were still trying to come to terms with the failure of the London Conference, the American Secretary of State, James Byrnes, sprang an unwelcome surprise on them. Byrnes, who had a reputation as something of a 'wheeler-dealer', was worried about the deterioration of relations at the Foreign Ministers Conference and believed that the differences could be resolved by another meeting. Without consulting the British he therefore sought an invitation from Stalin for a conference to be held in Moscow in December. The 'slippery Mr. Byrnes' was distrusted in Britain.[13] Apart from his reluctance to consult Britain in order not to be seen to be ganging up on the Soviet Union, he was suspected of being appeasement-minded. The idea of a Moscow Conference and Byrnes' reluctance to discuss common objectives with Britain before it began, was seen in the Foreign Office as further proof of the American Secretary of State's determination to seek an accommodation with Stalin over the head of the British.[14]

In practice the Conference proved more successful than the British believed it would be. This was largely because of the shock caused by the failure of the London Conference and a determination by all three states to avoid a showdown. The Conference resulted in what Victor Rothwell has described as 'an orgy of inessential concessions'.[15] Stalin accepted an Anglo-American investigation of undemocratic irregularities in Romania. All three agreed to the replacement of the Far Eastern Advisory Commission in Japan by a Far Eastern Commission consisting of the United States, the Soviet Union, the British Commonwealth and China (which, however, had no more authority than its predecessor). Agreement was also reached on a Peace Conference to be held in Paris to deal with the satellite states. As a result, France was granted the equality of status which Britain had fought for at the London Conference. The only issue which produced heated arguments and a continued deadlock was Iran.[16] Despite this, the British delegation felt that some progress had been made. Following a long talk with Stalin, Bevin believed that the Soviet leader had a clearer understanding of Britain's position. He went out of his way to emphasise 'the peaceful and (in the Middle East and India) progressive nature of British policy, and the need to respect the independence of Turkey'. At the same time he told Stalin that 'there was a limit beyond which we could not tolerate continued Soviet infiltration and undermining of our position'.[17]

Despite the relative success of the Moscow Conference, in the early months of 1946 British officials continued to ponder Soviet foreign-policy objectives. One of the most influential interpretations came from Frank Roberts, who was Minister at the British embassy in Moscow. Roberts was a personal friend of the American ambassador Walter Bedell-Smith, and a close associate of George Kennan, whose analyses of Soviet behaviour had such an important impact on American containment policy. Roberts held similar views to Kennan and his reports appear to have been as influential in London as Kennan's were in Washington.

In his early dispatches in 1946 Roberts warned that 'the Soviet Union is, and must be, fundamentally hostile to the outside capitalist world and, in particular to America and to a social democratic Britain'. Apart from ideological competition there was also the fact that Britain was seen in Moscow as a major obstacle to Soviet expansion around its borders. 'Its world empire with its main lines of communication running through the Mediterranean and the Middle East to India and Australasia' meant that British and Soviet interests were likely to clash as the Soviet leaders pursued unfulfilled Tsarist ambitions. Roberts argued that the Soviet Union was using 'Sudeten' techniques of posing as the protector of minorities in Turkey, Iran and Afghanistan as part of a campaign to expand Soviet influence into the Middle East.[18]

Roberts recommended a resolute response from Britain to keep the Soviet Union out of the Middle East. If this was not done, 'the temptation to penetrate an area divided by local, class and dynastic jealousies and to supplant a weakened and humiliated Britain might . . . prove irresistible, more particularly as just beyond the now debated territories of Persian Azerbaijan, Turkish Armenia and Kurdistan are to be found the Kirkuk oil fields and beyond them the oil fields of Kuwait, Qatar, Bahrain and Saudi Arabia'. The same opportunistic and relentless policy was evident in Eastern Europe, where the Soviet Union was intent on excluding all Western influence from the area and establishing control over the whole of Germany. Once again a determined response from Britain was needed.[19]

Roberts argued that it was not clear whether the reckless and even warmongering mood in Moscow was 'purely tactical' and how far it 'represented the first steps in a carefully considered long-term offensive strategy'. He was inclined to believe that the Soviet Union did not want war and would draw back from any risk of full-scale confrontation – especially with the United States. Nevertheless, the Soviet leaders appeared 'desperately' anxious to gain every 'advanced position' they could before international relations became less fluid. Roberts recommended that Britain should react firmly, develop close ties with the United States, without ostentation, and educate the general public to shed their illusions about the Soviet Union.[20]

Roberts' dispatches were well-received in the Foreign Office and provided the basis of British policy towards the Soviet Union in the early postwar period. Bevin was so impressed with the analyses that he ordered them to be shown to the Cabinet.[21] The Foreign Secretary, however, was still unwilling to throw down the gauntlet to the Soviet Union. He continued to retain the hope that, despite the aggressive nature of Soviet policies, some form of accommodation with the Soviet Union was still possible.[22]

This reluctance by the Foreign Secretary to abandon the quest for cooperation with Moscow became the source of some disquiet among Foreign Office officials in the second half of 1946 and in 1947. In April 1946 Warner, as head of the Russian section of the Foreign Office, produced a long memorandum on 'The Soviet Campaign against this Country and our Responses to it'. In contrast to his relatively favourable approach towards the Soviet Union in the later war years, Warner now argued that Russia was practising 'the most vicious power politics'. Recent Soviet policies were evidence that:

> the Soviet Union has announced to the world that it proposed to play an aggressive political role, while making an intensive drive to increase its own military and industrial strength. We should be very unwise not to take the Russians at their word, just as we should have been wise to take *Mein Kampf* at its face value.
>
> All Russia's activities in the past few months confirm this picture. In Eastern Europe, in the Balkans, in Persia, in Manchuria, in Korea, in her zone of Germany, and in the Security Council; in her support of Communist parties in foreign countries and Communist efforts to infiltrate Socialist parties and to combine left wing parties under Communist leadership; in the Soviet Union's foreign economic policy (her refusal to

> co-operate in international efforts at reconstruction and rehabilitation, while despoiling foreign countries in her sphere, harnessing them to the Soviet system, and at the same time posing as their only benefactor); in every word on foreign affairs that appears in the Soviet press and broadcasts . . . the Soviet Union's acts bear out the declarations of policy referred to above.[23]

Warner went on to argue that it could be assumed that the Soviet Union, in its present state, did not desire war. Nevertheless the danger of war remained. The Soviet use of military pressure in areas which affected British and American vital interests would require judgements about how far they could go without making war inevitable. This could lead to miscalculations (as in the case of Hitler and Poland) which might result in war.

Warner concluded that the 'Soviet ideological war' against 'liberal, democratic and Western conceptions' compelled Britain and other liberal democracies not only to defend themselves but also to conduct a worldwide campaign against Communism. The anti-Communist campaign should concentrate on exposing the myths which the Soviet government used to justify the policies. These myths included the idea that Germany was being built up to oppose Russia; that Russia alone gives disinterested support to colonial peoples; and that bad relations were the result of British aggressive designs against Russia. Warner also called for such a campaign to give moral and material support to all those struggling against Communism. In this respect Warner's memorandum foreshadowed the Truman Doctrine which followed a year later.

The ideas expressed in the memorandum were very influential and were shared by 'the highest in the land'.[24] In July the Prime Minister sent a query to the Foreign Office based on the assumption of a Soviet masterplan to strike at Western interests everywhere. Clement Attlee suggested that 'as Russian tactics in Europe and Asia follow the same pattern, it would be useful if our representatives in the East could be given early notice of tactics followed in the West and vice versa so that they would be forewarned'.[25]

The memorandum was produced just a month after Churchill's famous speech at Fulton, Missouri warning that an 'iron curtain' had descended across Europe. Although the Foreign Office regarded Churchill's speech as being unwelcome and premature, Bevin refused to repudiate it.[26] Nevertheless, it appears that the speech made him 'more determined than ever to practise the greatest patience towards the Soviet Union'.[27] The Foreign Secretary's ambivalence was shared by public opinion at the time. Within a few days of Churchill's speech Gallup found that 39 per cent disapproved of what he said while 34 per cent approved (16 per cent had no opinion). In

April 1946 a poll on the sources of Soviet behaviour revealed that 42 per cent believed that Soviet foreign policy was mainly the result of security considerations, while 26 per cent saw it as being inspired primarily by 'imperialist expansion' (32 per cent had no opinion).[28]

The Foreign Secretary shared many of the views about the Soviet Union expressed by Churchill and Warner. He also accepted the need, emphasised by both, that Britain must establish close ties with the United States. On 13 February 1946 Bevin wrote to Attlee emphasising the importance of the Anglo-American alliance:

> I believe that an entirely new approach is required, and that can only be based upon a very close understanding between ourselves and the Americans. My idea is that we should start with an integration of British and American armaments and an agreement restricting undesirable competition between our respective armament industries. The next step would be the adoption of parallel legislation in both countries to give their governments real control over the production of arms.[29]

Bevin, however, remained concerned about the volatile nature of American opinion and this contributed to his determination to seek 'the unpromising alternative of normal relations with Stalin's Russia until he was absolutely convinced that the quest was hopeless'.[30]

The Foreign Secretary's continuing interest in cooperation with the Soviet Union was shown in May, when a new British ambassador, Sir Maurice Peterson, was sent to Moscow with instructions to make a further effort to improve relations between the two countries. Peterson took with him a letter from Attlee to Stalin arguing that Britain was prepared to replace the wartime Anglo-Soviet Treaty with a new fifty-year alliance. The initiative, however, failed to improve the poor state of Anglo-Soviet relations. When Peterson met Stalin, the Soviet leader told him that he 'saw no good in prolonging the Treaty if account were taken of our present attitude towards Russia'.[31]

By the autumn, relations had deteriorated seriously. Roberts reported from Moscow in September that the attitudes of the Soviet leaders were 'so rigid' and their demands 'so ridiculously high' that he was becoming more and more worried about their peaceful intentions. A few weeks later the Moscow embassy reported that Soviet propaganda was warning that war with the Western powers was inevitable. Britain and the United States were portrayed as being the same as Hitler's Germany. Bevin told Attlee at the end of September that the behaviour of the Soviet delegates at the Paris Conference and the Security Council was wholly intransigent. They were 'attacking without reason'. Despite the conciliatory approach by Britain and the United States, the Soviet Union was pursuing 'a war of nerves all over the world'.[32]

At the beginning of 1947 the Foreign Office produced a major report designed to replace Orme Sargent's memorandum on 'Stock-Taking after VE-Day', written in the summer of 1945. The report, which was largely the work once again of C. F. A. Warner, described the hopes for great-power cooperation contained in Sargent's paper as 'illusory'.[33] It was argued that 'on VE-Day it might have been possible, though it was already difficult, to believe that the Soviet government intended to make a reality of Three Power collaboration'. This, however, was no longer possible. Although it was unlikely that the Soviet Union wanted war, they seemed bent on using their military power and the 'fifth column' of a world Communist movement 'to undermine British and US influence in all parts of the world and where possible to supplant it'. Warner, however, also remained concerned that war by miscalculation was a possibility: 'an extra bit of clumsiness and an outburst of mass emotion in America, coinciding with a sudden spasm of suspicion or display of truculence in Russia, might produce results which would get beyond control and lead to disaster'.[34]

The report concluded with a number of modifications to Sargent's 1945 memorandum which were needed in the light of Soviet behaviour. First of all, Sargent's proposal that British foreign policy should be based on the principle of Three Power cooperation had been undermined by intractable problems which had arisen in Eastern Europe and the Mediterranean. Such a course of action was now increasingly difficult to pursue. Sargent had also emphasised the need for an independent foreign policy. Warner argued that although this objective remained valid, Britain's economic weakness and Soviet threats, made 'too great independence of the US . . . a dangerous luxury'.[35]

Despite this gloomy assessment of Anglo-Soviet relations the Foreign Secretary was still unwilling in the first half of 1947 to abandon the principle of Three Power cooperation. Indeed he insisted on negotiations for a new Anglo-Soviet treaty, against the opposition of many of his top officials. In February Warner argued that the Soviet purpose in the treaty negotiations was 'to drive a wedge between Britain and the United States, to deflect Britain away from agreements with other countries and generally to weaken Bevin's position'.[36] Many officials appear to have hoped that the negotiations would fail. According to Gladwyn Jebb, success would be 'a little incongruous and perhaps even dangerous' if East–West relations in every other respect continued to deteriorate. With the negotiations dragging on in mid-1947, officials were still trying to persuade Bevin that the possibility of securing a treaty was not worth the concessions which would be necessary.[37]

The Foreign Secretary, however, continued to urge patience in dealing with the Soviet Union and castigated any loose talk about 'anti-Soviet groupings'. In his speech to the Labour Party Conference in May 1947 he emphasised the need for continued wariness towards Germany, but made no criticism whatever of Soviet behaviour. By mid-1947 many of his officials believed that he was 'excessively cautious' in standing up to the Soviet threat. Even when the treaty negotiations collapsed with the Soviet rejection of Marshall Aid, however, Bevin continued to tell his officials that he wanted 'a less anti-attitude' from them.[38]

On the surface it might appear that the Foreign Secretary's determination to keep open the option of collaboration with the Soviet Union was the result of the pressure exerted on the government by left-wing MPs during late 1946 and early 1947. Following its election in 1945 the Labour government was in a strong position in Parliamentary terms. It had captured 393 out of the 640 seats in the House compared with the Tories' 213. The government's commitment to a radical reform programme on the domestic front could be put through with little difficulty due to the overwhelming support given to it by Labour MPs. As Alan Bullock has argued,

> the implementation of the Beveridge Report (which has been described as the real 'social gospel' of the 1940s); the creation of the National Health Service and the welfare state; the commitment to full employment and economic planning; the nationalization of coal, steel, transport and the power industries; the transformation of the Empire by the grant of independence to India, Burma and Ceylon; the retention of controls, rationing and food subsidies; the wartime policy of 'fair shares' to regulate the transition from war to peace – all these could be viewed by the Left as the first stage in the introduction of socialism and by the majority of the Parliamentary Party and Labour voters as carrying out Labour's election programme pretty well to the letter.[39]

Labour's foreign and defence policy, however, was a different matter. Bevin's at times outspoken criticisms of the Soviet Union and his desire to achieve close ties with the United States were seen in some quarters of the Labour party as a betrayal of socialist principles. Increasingly in late 1946 and early 1947 Ernest Bevin was singled out for criticism by left-wing MPs in his own party. The campaign against him reached a high point in November 1946 when 57 Labour MPs tabled an Amendment to the Address complaining about his foreign policy. This represented a vote of censure by a sizeable proportion of the government's own backbench MPs. This was followed by the publication of *Keep Left* in April 1947, arguing for a socialist foreign policy and condemning the policies which had been followed since 1945. In March the government was also forced to retreat on the sensitive question of conscription when faced with a backbench revolt.

It seems unlikely, however, that these criticisms had any great influence over Bevin's attitude towards the Soviet Union. The Foreign Secretary was somewhat scornful of his critics and remained determined to follow the foreign policies he believed best served British interests. This is clearly illustrated by his attitude towards the External Affairs Group which was set up to provide an opportunity for backbenchers interested in foreign affairs to meet regularly with the Foreign Secretary to air their views. There were similar groups associated with other government ministries and the Ministers usually hand-picked the best MPs they could find and turned a potential source of trouble to their own advantage. In his memoirs Hugh Dalton, who was Chancellor of the Exchequer at the time, writes that:

> Bevin in sharp contrast had a terrible group on Foreign Affairs. He did not pick it, as I did. He let it pick itself. And in came all the pacifists, and fellow-travellers, pro-Russians and anti-Americans, and every sort of freak harboured in our majority. . . . The group as whole was hopeless. Bevin seldom met them. . . . They met, not in the Foreign Secretary's room, but in Committee Rooms upstairs. And there they drafted critical resolutions and prepared argumentative papers which, if ever he read them, merely infuriated him. One difference between Bevin and myself in 1945 was that I knew a majority of Labour MPs and made it my business to get to know them all, whereas Bevin knew very few of them and made no serious effort to extend his knowledge.[40]

By temperament and personality, therefore, Bevin was unlikely to be swayed by disillusioned left-wing MPs, uncomfortable as their criticisms often were for the Foreign Secretary from within his own party.

Similarly, despite the feeling within the Foreign Office in early 1947 that he was overly cautious in his approach towards the Soviet Union, Bevin insisted on pursuing the policies he felt were most appropriate. Bevin was his own man on these matters. Despite the difficulties with the Soviet Union and his willingness to take a tough line when he thought it necessary, he continued to work for a resolution of the disagreements and refused to accept that relations had deteriorated beyond the point of no return. Cooperation with the Soviet leaders might be difficult and indeed unlikely but for the moment he had not given up hope.

Contrary to the views of many revisionist historians, therefore, British policy towards the Soviet Union in 1946 and early 1947 was not characterised by a single-minded determination to abandon Three Power collaboration and an anti-Communist crusade by the Foreign Secretary. Bevin certainly spoke his mind at the Foreign Ministers conferences and major disagreements arose over Eastern Europe, Germany and the Middle East. Foreign Office officials in London and Moscow were also highly suspicious of Soviet motives and many of them were increasingly convinced that there was little to be gained by trying to maintain Three Power cooperation. Although the Foreign Secretary shared much of their anxiety about Soviet behaviour, especially the expansion of Soviet control over the Eastern European states, he nevertheless remained determined to try and work for an accommodation with Moscow while stoutly defending British interests. It was in this context that Bevin pursued his ideas for a Western European group and a close relationship with France.

4 Towards a Treaty with France

Bevin inherited from Anthony Eden the scheme for a Western European Group, which was to have an Anglo-French alliance as its foundation. Shortly after arriving at the Foreign Office he was presented with two memoranda arguing that 'a close Anglo-French alliance is very much in the interests of this country'.[1] The first pointed to the difficulties of making progress towards such a treaty which centred on the problem of the Levant and the western frontier of Germany. The second provided a detailed history of the Western bloc scheme during the war and contended that the arguments in favour were as strong as ever.[2]

At this time there was a strong sense of frustration amongst various officials in the Foreign Office, who believed that sound Anglo-French relations were crucial for Britain's future policy towards Western Europe.[3] By the end of the war the relationship between the two countries had deteriorated sharply. De Gaulle had made it clear that no progress could be made towards an Anglo-French treaty until the Levant and German frontier problems had been resolved. This meant British acceptance that France should retain its prewar position in the Levant and a delineation of the western boundaries of Germany which excluded both the Rhineland and the Ruhr.

The position, however, had been made even worse by the 'provocative' action of the French in reinforcing their garrison in Beirut in April 1945. This led to anti-French disturbances, which caused the French to bombard Damascus at the end of May. Britain's decision to use troops to keep the various factions apart only served to confirm de Gaulle's suspicions that the British had ambitions to replace France in the region.[4] De Gaulle's hostile reaction was summed up in a comment to Duff Cooper on 4 June when he said: 'we are not, I recognise, in a state to wage war on you now. But you have outraged France and betrayed the West. That cannot be forgotten.'[5]

Despite this unpromising atmosphere, when Bevin arrived in the Foreign Office he was determined to break with the recent past and to overcome the personal antagonism between de Gaulle and Churchill which had blighted relations at the end of the war. In a series of meetings in the Foreign Office between 10 and 17 August 1945 he emphasised the need to put relations between Britain and France on a better footing as soon as possible so that their friendship could be the foundation for greater collaboration between the states of Western Europe as a whole.[6]

In talks with members of the Western Department on 13 August Bevin argued that his long-term aim was 'extensive political, economic and military co-operation throughout Western Europe, with an Anglo-French alliance as a cornerstone'.[7] He expanded the geographical scope of the projected Western European group to cover the countries on the Mediterranean and Atlantic seaboards, including Greece, Italy, France, Belgium, the Netherlands and Scandinavia. This being so, 'Bevin was not simply treading a path cleared for him but rather tapping a fund of ideas of the desirability of European co-operation – especially on the economic level – which had been in his mind during the war and in the years before'.[8] Bevin's plan was immediately dubbed 'the grand design' in the Foreign Office.[9]

Bevin's emphasis on financial and commercial cooperation between Western European states at this time seems, in part at least, to have been due to a desire to achieve economic independence from the United States. Bevin hoped for a continuation of good Anglo-American relations in the postwar period but he was not convinced that it would be possible to prevent the United States from returning to isolationism. It was therefore important that Britain should be able to stand more on its own feet especially in economic terms. The experience of the immediate postwar period was not auspicious. In early August 1945 Truman had cut off all lend-lease aid to Britain, creating what Lord Keynes described as 'an economic Dunkirk'. The attempt by the Labour government to secure an interest-free loan of $6.6 billion to deal with the economic crisis was rebuffed by Washington and, after difficult negotiations, a loan of $4.4 billion at 2 per cent interest was agreed. The friction caused by these events led *The Economist* to complain that:

> It is aggravating to find that our reward for losing a quarter of our national wealth in the common cause is to pay tribute for half a century to those who have been enriched by the war. . . . Beggars cannot be choosers. But they can by long tradition put a curse on the ambitious and the rich.[10]

This view was reinforced by the events which took place at the Council of Foreign Ministers meeting between 11 September and 2 October. The conference was characterised by a serious disagreement between the United States and Britain, on the one hand, and the Soviet Union, on the other, over Soviet actions in Bulgaria and Romania. A Soviet request for the trusteeship of the former Italian colony of Tripolitania was also met with fierce resistance by Britain and the United States. Concerted policies in these areas, however, did not mean that relations between the two countries were wholly amicable. The problem lay largely in the difficult personal relations between the American Secretary of State, James Byrnes, and Ernest Bevin. The Americans had not taken to Bevin at the Potsdam conference. He was regarded as 'graceless and rough'.[11] Byrnes, on the other hand, was not well-regarded in the Foreign Office in London. The Permanent Under-Secretary, Sir Alexander Cadogan, thought he was 'really rather a disaster; and it's very difficult to work with him. He appears shallow and impulsive – and rather slippery.'[12] Bevin also formed a distrust of Byrnes at the outset of their relationship, which never really improved. This personal antipathy between the Foreign Secretary and Secretary of State left the British feeling 'isolated and alienated from the uncommunicative Americans'. During the conference Bevin informed his colleagues in the Cabinet that the British delegation 'were faced with increasing hostility between the United States and Soviet delegations each of whom sought to strengthen its own position without regard to our point of view'.[13]

Bevin's immediate reaction to the unsatisfactory conclusion of the Foreign Ministers meeting was to suggest to Georges Bidault, the French Foreign Minister, that there should be an Anglo-French treaty. In many ways this was a good moment for such a treaty. The two outstanding issues which had so far prevented negotiations appeared in the autumn of 1945 to be near resolution. Bidault's presence in London for the Council of Foreign Ministers meeting allowed Bevin to initiate conversations on the Levant issue which were to produce an acceptable settlement for both by the end of the year.[14]

There was also reason to be hopeful over a resolution of the German frontier problem. The French had consistently demanded that the Ruhr should be placed under international control and detached from German administration. This proposal had not been supported in the Foreign Office because of the longstanding fear that the dismemberment of Germany would give rise to the resurgence in the future of a 'vengeful, nationalist Germany'.[15] Foreign Office officials, however, were persuaded to moderate their view on the issue by Bevin, who was not wholly opposed to the French view. As Sean Greenwood has argued, he had come to the Foreign Office with a history as a hard-liner and was by no means averse to further amputation of Germany territory'.[16] Since coming to office he had been interested in the idea of severing the industrial Ruhr from the rest of Germany and putting it under international control. In August he had told members of the Western Department that 'his long-term aim was to make the Ruhr industries a central pivot in the economy of an eventual "Western Union"'.[17]

Although Bevin emphasised to officials in December that it was important 'to work steadily towards the closest cooperation and integration economically, socially and militarily with our Western neighbours', at this stage it was economic cooperation which appeared to be at the forefront of his mind.[18] This was not altogether surprising, given Britain's desperate economic circumstances in the immediate postwar period. The Foreign Secretary seems to have viewed the West European group, 'with its economic recovery supercharged by an internationalized Ruhr', as one of 'a series of European economic groupings which would serve to regenerate the continent and preserve peace'.[19]

Although Bevin had not decided at this stage whether the Ruhr should be detached from Germany, his views were close to those of the French. As a result the prospects for an Anglo-French treaty seemed bright. At the end of the year Bevin even went as far as to order the Foreign Office to prepare a draft treaty.[20]

By the end of January 1946, however, the whole situation had changed as a result of political development in France. In October 1945 the General Election had resulted in the Communist Party becoming the largest group in the Assembly. It took a further month before de Gaulle was able to construct a governing coalition. This in itself had little effect on Anglo-French relations. On 20 January 1946, however, only two months after forming this government, de Gaulle resigned precipitating a new political crisis. This raised deep anxiety in Britain that the Communists, the Mouvement Républicain Populaire and the Socialists would not be able to work together effectively and France would slide towards some form of dictatorship of the Right or the Left.

Bevin regarded these events in France with the greatest seriousness. In his view, civil war was a distinct possibility within a year and the Sovietisation of France could now not be ruled out.[21] This deep anxiety about domestic events in France was reinforced by the deterioration of relations with the Soviet Union in early 1946. The emerging crisis over Iran, Soviet pressure on Turkey and the consolidation of Communist control in the Soviet zone in Germany, all fuelled growing worries about the continuation of Big Three cooperation. These events in turn coloured perceptions of domestic circumstances in France and helped create 'a state of panic' in the Foreign Office.[22]

In early February the Foreign Office accepted that this interpretation of the political crisis in France had been somewhat exaggerated. It was now clear that the tripartite government was working satisfactorily without de Gaulle.[23] The fears of civil war had proved to be largely unfounded. Nevertheless, 'the profound crisis of confidence in the vigour of the French political system which beset the Foreign Office in late January could not easily be erased and was, henceforward, to condition all British consideration towards France'.[24] Anxiety about the stability of the French political system remained an important consideration in the British attitude towards the French in the months that followed.

The most immediate result of the events of late January was that France was given a diminished emphasis in British policy towards Western Europe. Greater attention was now given to the western sectors of Germany. This change of priority was evident in a paper written by Hasler, the Head of the Supply and Relief Department of the Foreign Office on 27 March.[25] Hasler argued that 'if we are to have a European policy which makes sense to Europeans and is not a matter of passing likes and dislikes for particular countries, the secret lies not in France but in West Germany'.[26] By this time Bevin had decided not to accept French demands for a wholly independent Ruhr and to focus on fusing the American and British zones in Germany. These decisions reflected 'the devaluation of France as the centre of British interest in Western Europe'.[27]

This change of priority did not mean, however, that Britain had wholly lost interest in an Anglo-French treaty. The Deputy Under-Secretary in the Foreign Office, Sir Oliver Harvey, reasserted the case for a treaty in mid-February, justifying it in terms of the need to reinforce 'Britain's friends in France' who were opposed to Communism.[28] In February and March a series of meetings were held in the Foreign Office, designed to 'clear the minds' of officials after the events of late January and establish a more coherent policy towards future relations with France and a Western European group.[29] On 13 March Bevin told officials that despite the problems he remained determined to continue to work steadily 'towards the closest co-operation and integration economically, socially and militarily with our Western neighbours, without at this stage creating any formal regional group. In accordance with this policy we are trying hard to get our relations with France cleared up first.'[30]

At the same time that the British were trying to sort out how they should proceed on these issues, a new administration in France was also beginning to give some attention to the question of some form of Anglo-French treaty on the lines of the Franco-Soviet Pact of December 1944. On 30 March 1946 the President of the Council, M. Gouin, made an important speech which suggested that a significant modification had taken place in the attitude of the new French government which had replaced de Gaulle.[31] M. Gouin indicated that it might now be possible to conclude a treaty between the two countries 'at once without waiting for a settlement of the Western frontier'.[32]

As a result of Gouin's speech, Sir Oliver Harvey was dispatched to Paris to visit the French Foreign Minister, M. Bidault, and M. Gouin to see what they had in mind. Harvey's visit, however, proved to be somewhat premature. In his talks with Bidault on 4 April he was told that the matter was still being considered by the French government but that Britain could expect a statement after a government meeting on 9 April.[33] Bidault did, however, suggest to Harvey that he thought France would be willing to enter into negotiations with the UK 'to discuss all outstanding questions with a view to arriving at an alliance as soon as possible'. The French Foreign Minister explained that he was obliged to say 'all outstanding questions' so as not to give the impression of a change of policy, although in fact the government was willing to form an alliance without the Ruhr and Rhineland questions being settled.[34]

This, however, was still rather too vague for the Foreign Office in Britain and for the Foreign Secretary, Ernest Bevin. Discussion in the Foreign Office on the Harvey visit raised continuing doubts over whether the French had in fact changed their minds on linking an alliance to outstanding issues. From Britain's point of view, it remained important that the two questions should be kept completely separate.[35] 'Either the alliance was right or wrong for the interests of the two countries.'[36] As a result, Bevin sent a telegram to his ambassador in Paris, Duff Cooper, urging him to take no action. To Bevin the ball was still very much in the French court.

British scepticism proved to be correct. Despite the statements by Gouin and Bidault indicating a change in position, domestic preoccupations in France precluded any definite decision on pushing for a treaty with Britain for the remaining part of the year. In consequence, from the standpoint of the British government, France remained officially committed to solving the German question before negotiating an alliance, thus precluding any further British initiatives for the time being.

Discussions on a French pact and a Western European group did, nevertheless, continue to preoccupy officials within the Foreign Office during the remaining months of 1946. Much of the debate during this period tended to centre on a memorandum produced by Sir Nigel Ronald who, as head of the wartime Economic and Reconstruction Department, had been dealing with these issues on and off since 1942.[37] In his memorandum produced in December 1945 and circulated in the early months of 1946, Ronald (who was Acting Assistant Under-Secretary in the Foreign Office) argued that an Anglo-French Treaty was the keystone of any Western European defence system.[38] He urged the negotiation of a pact with France which would be followed by the establishment of a group consisting of 'Belgium, Luxembourg, the Netherlands, Norway, Denmark, the British zone of Germany and eventually Portugal and Spain'.[39] Ronald argued that Britain must take the lead in this process, to give confidence to the minor Western allies in order to prevent them falling under Russian influence and to secure the long-term containment of Germany.

In his memorandum, Sir Nigel was attempting to link an Anglo-French alliance, the containment of Germany and the organisation of international security. In this sense his proposals were not confined simply to the combination of Western European states. He argued for other regional groups as well, including an Eastern European Group organised by the Soviet Union, all of which would make up a world-wide security jigsaw under the auspices of the United Nations.[40]

All of the relevant departments in the Foreign Office and the Chiefs of Staff commented on the Ronald memorandum and on 5 December 1946 a meeting was held in his room in the Foreign Office to discuss the various proposals in more detail.[41] At the meeting a great deal of scepticism was expressed by various officials about the plan. It was argued that there were two alternative aspects under which Ronald's scheme could be considered. Firstly, as a method of enabling Britain to create a system of defence in Western Europe against a possible Russian attack. And secondly, as a means of improving Britain's relations with Russia and thereby diminishing the likelihood of such an attack.

As far as the first aspect was concerned, the CoS argued that a regional defence system in Western Europe offered little or no attractions as a method of defence against a possible Russian attack. Without the participation of the United States, they argued, there was virtually no chance of the countries concerned being able to withstand a Soviet invasion.

Attention was therefore given to the plan as a method of improving relations with the Soviet Union which had been steadily deteriorating since the end of the war.[42] From this point of view it was emphasised that it would be necessary to remove those features of the proposals which might make the Soviet Union think it was directed against them. It was pointed out, however, that if this was done the plan might then become unacceptable to the Americans. There was widespread agreement at the meeting that the 'Americans were in the last resort, our only hope of defence against Russia and it was therefore axiomatic that we must at all costs avoid the risk of antagonizing them'.[43] It was felt, therefore, that any plan for a Western European group which was devised must be concerned with taking the United States along with Britain.

The conclusion of the meeting was that at first sight it seemed extremely difficult to devise a scheme of the kind Sir Nigel Ronald had in mind which would both be acceptable to the Americans and would also be calculated to achieve its avowed objective, namely the improvement of relations with Russia. It was generally agreed that much more thought needed to be given to the proposals.

The Ronald plan, however, was circulated to various interested ambassadors for their comments. The most critical of the replies came from Duff Cooper in Paris.[44] Cooper had unavailingly urged the need for a British continental commitment in the 1930s and in 1944 had written a long memorandum to Anthony Eden on the need for a postwar alliance with France and the creation of a Western European grouping.[45] He welcomed this aspect of the Ronald proposals. His opposition to the plan, however, centred on two issues. Firstly, he believed that the idea of linking the plan so closely with the United Nations and allowing the Security Council to consider the proposals would result in paralysis. The whole idea, Cooper argued, would get 'bogged down' in endless discussion and would come to nothing at a time of growing tension between the Big Three Powers. Secondly, he was very critical of the cautiousness expressed in the proposals and the constant references to the need 'to take the Americans along with us'. Unlike the CoS, Cooper saw Britain's future not in close association with the United States but in partnership with Western European states. He therefore argued that if Britain made 'every move contingent on American prior approval' it would not be possible to attain what he saw as the real objective of putting Britain 'more nearly on a par with the USA and USSR by close association with our neighbours'. Duff Cooper's dream remained, as it was in 1944, a federation of the Western seaboard of Europe together with the principal powers of the Mediterranean. Such a federation, he believed, could become the strongest of the three great world power combines – 'an alliance so mighty that no power on earth would have dared to challenge it'.[46] To achieve this, he argued that at least Ronald was right to stress the importance of an alliance with France as a starting point.

As a result of these and other criticisms, it is clear that by mid-December 1946 the Ronald plan was 'in the process of undergoing radical transformation' in the Foreign Office.[47] Nevertheless, a great deal of uncertainty still existed in British circles over how to proceed. On 21 December Sir Orme Sargent, now Permanent Under-Secretary of State in the Foreign Office, sent a minute to the Secretary of State in which he pointed out that after the political difficulties of the past few months in France, a new government would soon be in office.[48] This, he felt, might provide the opportunity for a fresh approach to the question of an Anglo-French alliance. It is interesting to note that at this stage, despite his general support for the idea of closer cooperation with France and other West European states, Bevin remained dubious, both in public and in private, about a Western European group and cautious about pushing hard for an Anglo-French Treaty.[49] Twelve months later he was to move decisively towards the formation of a Western European security system, but in late December 1946 he expressed himself 'in favour of waiting a little longer'.[50] From the Foreign Secretary's point of view there was the problem that M. Blum's new government would probably only remain in office for a few weeks and it might be just as well, therefore, to wait for a more permanent government to be formed. There were also the indications that the French still remained committed to solving the 'German problem', which restricted the Foreign Secretary's room for manoeuvre.

Bevin's attitude seems to have been radically changed by M. Blum's letter to the Foreign Secretary which arrived on 1 January 1947.[51] In his letter Léon Blum declared that 'nothing would give me greater joy or make me prouder than during my brief stay in office to be able to set my signature' to a treaty between Britain and France. The crucial part of the French socialist leader's message to Ernest Bevin was that he sought no prior solution of outstanding problems and wished to follow up his initiative as soon as possible. He did, however, feel obliged to warn the British that a treaty would only be welcome with complete satisfaction in France if the public could 'look forward at the same time to the early solution of a problem of whose vital importance they are well aware'.[52] Despite this qualification the final stumbling-block which had bedevilled Anglo-French relations since the end of the war had been swept away by Blum's initiative. Arrangements were rapidly made for him to visit London on 13 January to discuss the question of an Anglo-French treaty. When he arrived, however, it became clear to the British that his major preoccupation was with domestic considerations in France. He used his visit to appeal to his hosts to send more coal to France, which was desperately needed. As a result the question of a treaty became embroiled in important, but rather mundane, discussions about whether Britain could respond to Blum's appeal.

For the British government, Blum's plea created something of a dilemma. The onset of the worst winter in living memory meant that Britain had little coal to offer. At the same time there was concern that Blum's position at home should not be undermined and that everything should be done to prevent France from slipping into the Soviet orbit.[53] In the negotiations which took place on 13 and 14 January, however, Bevin was only able to offer Blum an early restoration in coal allocations when the circumstances permitted.

It was in this context of Britain's failure to provide immediate coal supplies to France that Bevin and Attlee decided to offer Blum an alliance. For the British Foreign Secretary, the offer of a treaty 'contained elements of both expediency and of high politics'.[54] Should Blum and the socialists remain influential, the longer-term objectives of a wider Western European group could be pursued further. If not, then 'the suggested alliance would at least save Blum from embarrassment, might bolster the flagging electoral performance of the Socialists and could possibly keep France in the Western camp'. Bevin undoubtedly still harboured the broader vision of greater European cooperation based on an Anglo-French alliance and the meeting with Blum helped to rekindle this design. Nevertheless, there was a strong element of opportunism in Bevin's offer to Blum. Immediate objectives reinforced the pursuit of the 'grand design'.[55]

On 16 January, at the end of Blum's visit to London, a communiqué was issued, declaring the intention of both countries to negotiate an alliance. There were still, however, some disagreements to be resolved in the British camp before negotiations with the French could begin. The differences tended to centre on the kind of treaty that should be negotiated and the impact of the treaty on the Soviet Union and the United States. In particular, the decision by the Foreign Office to use the Anglo-Soviet Treaty of 1942 as a model for the Anglo-French Treaty after the 16 January communiqué was immediately questioned by Duff Cooper in Paris.[56] In a telegraph to Bevin on 18 January, Cooper suggested that a closer understanding was necessary with France than with Russia. He recommended, therefore, a general treaty which would include a commitment by the two powers to set up bodies of experts to negotiate more detailed agreements in various spheres, including commercial, colonial and defensive arrangements.

The ambassador's suggestions, however, caused some irritation within the Foreign Office itself. One official commented critically that this was 'yet another repetition of his familiar theme that we should go all out for putting teeth into the Alliance and that instead of being content to get gradually closer to the French by unofficial talks about military matters, colonial affairs, economic problems etc., we should supplement the Alliance by a series of public official statements about Staff Talks, colonial cooperation etc.'.[57] It was generally agreed within the Foreign Office itself that this would be a mistake and that it would be better to go 'slowly and quietly' on the matter.

In his reply to Duff Cooper the Foreign Secretary pointed out that he believed it was necessary to keep to the original approach and to model the treaty on the Anglo-Soviet and Franco-Soviet pattern, directed specifically against German aggression.[58] He rejected any wider understanding between the two countries because 'any suggestion that the treaty was not directed primarily against Germany would arouse suspicion and perhaps opposition not only in Moscow but in some other quarters including, I would imagine, the Communist party in France and their sympathisers in Belgium and elsewhere'.[59] Bevin feared that this would create the risk of losing the treaty altogether.

Apart from this concern over the reactions of the Soviet Union at this diplomatically delicate time, there were also those in Britain who were anxious about the impact of a treaty of this kind on Anglo-American relations. Those who had thought for some time that such a treaty would encourage isolationist feeling in the United States returned to the theme in January and February 1947. The Chiefs of Staff, in particular, were only prepared to make their approval of the alliance conditional on the assumption that it in no way impaired Britain's relations with the United States. From the viewpoint of the military chiefs, despite certain difficulties in Atlantic relations (associated with the loan agreement and Palestine) continued collaboration with the Americans was the keystone on which Britain's major strategy and planning was being based. For this reason, any risk of impairing those relations would be 'too great a price to pay for such a treaty'.[60]

The Foreign Office rejected such anxieties. They remained quite confident that such a treaty would not harm Anglo-American relations. Indeed it was argued that, on the contrary, the Americans were extremely pleased that, given some of the strains in Anglo-French relations in previous years, Britain and France were willing to cooperate more closely.[61]

Despite these differences of opinion in British circles the negotiations (which followed the 16 January communiqué) with the new French government of Paul Ramadier proceeded remarkably smoothly and quickly. The only real disagreement which arose was over the strength of the words to be used to describe the German threat. The French insisted that the treaty should be directed against the 'menace' of renewed German aggression.[62] Britain accepted the designation of Germany as the major threat to peace, but for various reasons wished to omit the word 'menace' from the treaty. In part this may have been due to a growing feeling within the Foreign Office that Germany, or part of that country, might in the longer term have to be brought into a future Western European system, particularly given the uncertainty of future relations with the Soviet Union. It was also due, however, to a very real concern not to undermine the proposed Byrnes Four Power Treaty which still lay on the negotiating table (despite the Soviet rejection of it in July 1946). There was a strong feeling in Britain that such references to Germany might make the Byrnes Treaty appear superfluous and encourage isolationist elements in the US Senate to oppose it. This posed the danger that Britain and the other Western European states might lose any chance of an American guarantee for their security which was considered so vitally important, especially by the British Chiefs of Staff.

After a series of meetings in mid-February, however, between Sir Orme Sargent and M. Massigli, the French ambassador in London, a compromise was worked out involving the acceptance of a new article in the treaty which contained the two phrases 'without prejudice to the Four Power Treaty' and 'prevent Germany from becoming a menace again'.[63] Thus the French achieved the strong wording they required, reflecting their feelings about Germany, and Britain secured a reference to their continuing interest in the Byrnes Treaty, reflecting their sensitivity over American opinion.

This achieved, the governments of both countries approved the Treaty in late February 1947. Both Britain and France were anxious to sign the Treaty before the Moscow Conference which was due to begin on 10 March.[64] Consequently on 4 March Bevin travelled to Dunkirk on his way to Moscow to put his signature, together with that of M. Bidault, to the final document which, from a British point of view, represented the successful completion of a policy initiated many years before.[65] For the past five years a series of proposals produced by the Foreign Office had all advocated the need to coordinate the policies of the two most important West European states in order to create the starting-point for some form of a Western European security system. The Treaty of Dunkirk finally achieved this objective.

Despite some of the disappointments with, and limited achievements of, the Dunkirk Treaty, it did prove to be important from the British perspective.[66] It helped to build up confidence in France, encourage the small West European states and demonstrated to the United States that a start had been made by the two most important states in Western Europe on working together. As such, the Dunkirk Treaty played a significant role in preparing the ground for the Brussels Pact in March 1948 which, in turn, also played an important part in encouraging the United States to participate in the wider Atlantic Alliance, binding the states of Western Europe and North America together.

It must be said, however, that at the time that the Dunkirk Treaty was signed there was not a widespread consensus, either in the Foreign Office or in British political circles generally, that the ultimate objective should be a *formal* West European group – or a wider security alliance involving the United States and directed against the Soviet Union. Certainly there were some who thought in these specific terms[67] and there were others who advocated closer European cooperation and an intimate partnership with the United States.[68] There were many different ideas, however, on how these objectives might be achieved, just as there were differing attitudes towards the significance of deteriorating relations with the Soviet Union at this time.[69] Uncertainty still existed over whether the Soviet Union or Germany was the more serious threat to British security and the peace of the world.

John Young has argued that the Dunkirk Treaty was 'a product of a distinct era lasting from 1944 to 1947, an era in which it had been hoped to build world peace on the Big Three and the United Nations with security measures directed against Germany'.[70] In another respect, however, it represented the beginning of a new era. Although Ernest Bevin was to tell the British ambassador in France as late as March 1948 that Britain could not 'afford to ignore the German danger', the Dunkirk Treaty nevertheless was the precursor of the anti-Soviet security arrangements which were to follow.[71] Just as the Treaty was signed, the immediate postwar era was coming to an end. Big Three solidarity was breaking up and the Soviet Union was increasingly perceived in Britain as a serious threat. The Dunkirk Treaty came to be seen as an important cornerstone in the search for greater political economic and military cooperation between Western European states which Bevin had championed in his August 1945 'grand design'.

This is not to suggest that Ernest Bevin had a blueprint that he implemented step by step. As we have just seen, the Dunkirk Treaty was as much the result of expediency as foresight on Bevin's part.[72] Nor is it to argue that at this stage the Foreign Secretary saw some form of Atlantic community as the next phase after his objective of European cooperation was achieved. In fact, from the evidence available Bevin seems to have been tempted for a time by Duff Cooper's idea of Britain playing a more independent role between the United States and the Soviet Union. In November 1945 after the unsatisfactory conclusion to the London Council of Foreign Ministers meeting he told the House of Commons that he could not 'accept the view that all policy and the policy of HMG must be based entirely on the Big Three'.[73] In late November and early December he also tried to raise with the Americans the idea of 'three Monroe areas'. The concept was never spelled out in detail but it seemed to be based on the idea that the United States, the Soviet Union and Britain should each have their own spheres of influence along the lines of America's traditional Monroe Doctrine.[74]

This does not mean that in the immediate postwar period Bevin had arrived at a clear view that Britain should pursue a middle path between American capitalism and Soviet Communism. In 1945–46 the way ahead seemed very uncertain and the Foreign Secretary was searching for some kind of order to allow Britain, as the weakest of the Big Three Powers, to pursue its national interests most effectively. In this context the attempt to achieve greater cooperation with Western European states, starting with

France, provided an important framework for British foreign policy. The Dunkirk Treaty represented a significant milestone in the pursuit of this policy. By the summer of 1947, as Anthony Adamthwaite has argued, Britain was playing an important role in the leadership of Western Europe. Indeed 'a rush of events was propelling her into Europe' at this time.[75] The independence of India, Burma and Ceylon, difficulties over continuing to shoulder the heavy responsibilities for Greece, Turkey and Palestine, together with the growing tensions of the Cold War, all led Britain to focus more sharply on helping to draw Western Europe closer together.

The Dunkirk Treaty represented a triumph for those officials in the Foreign Office who had advocated a treaty with France as the cornerstone of a Western European group ever since 1944. It would be wrong, however, to suggest that the treaty reflected a settled course in British foreign policy at this time. In practice, ambiguities and uncertainties abounded about the Russians and the Americans as well as about the Germans and the French. Despite the Dunkirk Treaty the major dispute between the Chiefs of Staff and the Foreign Office which had been at the heart of wartime planning remained unresolved. While the Foreign Office in general retained its preference for British leadership in Western Europe, the Chiefs of Staff remained convinced that Atlantic relations continued to be of greater importance. As we saw earlier, the CoS were only interested in the treaty with France provided that it did not undermine relations with the United States.[76] Some Foreign Office officials, however, were quite prepared to go along with the treaty irrespective of American opinion.[77] For the moment, the Dunkirk Treaty represented the ascendancy of the European option and political considerations over military options. It was not long, however, before the Atlantic option and a greater focus on security considerations were to move to the centre of the stage.

5 The Western Union and the Brussels Pact

The signing of the Dunkirk Treaty in March 1947 came at a time when Britain was facing very severe economic difficulties which had important implications for her foreign and defence policies. Despite the need to go on shouldering the burden of occupation duties, especially as relations with the Soviet Union deteriorated, economic pressures increasingly built up on the defence budget. The government's dilemma was highlighted throughout 1946 and 1947 with the Chancellor of the Exchequer, Hugh Dalton, forecasting economic disaster if defence expenditure was not reduced, while Ernest Bevin was urging his colleagues in the Cabinet to provide the necessary military muscle to back up the pursuit of foreign policy in an increasingly dangerous world.[1]

Despite the Foreign Secretary's interest in pursuing an independent role in the longer term he recognised that good relations with the United States would be of major importance in seeing Britain through its immediate difficulties. As we saw earlier, on 13 February 1946 he told Attlee that he believed a new approach was required between Britain and the United States.[2] A close relationship was of major importance to Britain. Despite the tough conditions laid down by the Americans for a loan to Britain in October 1945, Bevin argued in favour of an agreement, against some of his Cabinet colleagues. In his view it was essential to get help from the Americans 'if we are to rebuild our industry and restore our trade with Europe'.[3] In his view American assistance was also vital to ease the growing pressures on defence policy. The most dramatic example of this came in February 1947 when the British government sent a message to Washington, warning that unless the United States agreed to give financial help for the Greek armed forces, Britain would no longer be able to carry the burden of supporting the Greek government against the Communist guerrillas. Bevin's ultimatum worked and on 12 March Truman announced his 'doctrine' of supporting 'free peoples who are resisting attempted subjugation by armed minorities or by outside pressures'.[4]

For the British government the American help to Greece and Turkey which resulted from the Truman Doctrine was just as important in symbolic as in material terms. For the first time, despite all the uncertainties and difficulties with the United States over atomic energy collaboration, loan negotiations and Palestine, the United States had committed itself to intervening in the defence of Western European interests.

Another milestone in the shift of emphasis in British policy came with General Marshall's speech at Harvard on 5 June 1947 offering US economic aid to European countries. Shortly before the speech Bevin had concluded that the time had come to follow up the Dunkirk Treaty with a series of similar treaties with Belgium and Holland.[5] The promise of Marshall Aid, however, led to a postponement of Bevin's proposed approach to the smaller West European states. The moment was now 'inopportune' for such a political initiative.[6] Events would have to wait the outcome of the Conference of Sixteen Nations in Paris. When the Belgian government pressed Britain in July to pursue the matter of a treaty further, Bevin argued that attention for the moment had to be focused 'on the economic plane'.[7]

To Bevin, Marshall's speech of 5 June was of crucial importance, not only to Britain but to the states of Western Europe as a whole. The main threat facing Western Europe, he believed, was steady economic deterioration with the prospects of large-scale starvation, declining confidence in local currencies and even, perhaps, revolution. Will Clayton, the American Under-Secretary of State for Economic Affairs, warned the State Department at the end of May that 'without further prompt and substantial aid from the USA, economic, social and political disintegration will overwhelm Europe'.[8] For Bevin, Marshall's implied offer of support in his Harvard speech was therefore 'like a life-line to sinking men'.[9] It was a means by which the 'looming shadow of catastrophe' might be averted, and he moved swiftly to coordinate the West European response.[10]

Having played a major role (together with M. Bidault, the French Foreign Secretary) in helping to organise the West European response to the Marshall Plan, Bevin once again turned his mind to the wider dimensions of Western security. From 1945 to the spring of 1947, despite the emerging difficulties with the Soviet Union, Bevin had gone on searching for a satisfactory peace settlement in agreement with the Russians and the Americans. By the end of the Moscow Council of Foreign Ministers meeting in April 1947, however, Bevin was no longer prepared to allow the constant Soviet vetoes to maintain a paralysis of Western policy. He continued to hope that the Soviet Union would adopt a more reasonable approach to negotiations right up to the London Council of Foreign Ministers in November and December.[11] From the spring and summer of 1947, however, he gradually lost faith in such hopes. In September, Bevin had talks with M. Ramadier, the French President of the Council, in which he argued that the recent Russian moves had led him to the conclusion that the division of

Europe into Eastern and Western camps was now inevitable and 'it therefore became necessary to attempt to organise the Western states into a coherent unit. The time had now come for a return to the political plane.'[12] Bevin was not alone in thinking that the Western states ought to organise themselves more effectively at this time. The Canadian Foreign Secretary, Louis St Laurent, made an important speech on 18 September arguing that the time might be ripe for the creation of some form of Western security association.[13] St Laurent's speech was the first public suggestion by a member of a Western government that such an association should be considered. Bevin's however, was the first major initiative.

Bevin chose the collapse of the Council of Foreign Ministers meeting in London in December 1947 as the occasion on which to launch his plans for an organisation of West European states. On the morning of 17 December he had talks with M. Bidault, in which he followed up some of the things he had said to Ramadier in September. The time had come, he argued, to create 'some sort of federation in Western Europe, preferably of an unwritten flexible kind'.[14] Such an undertaking would require American assistance, Bevin told Bidault, and care would have to be taken in dealing with the Americans. It would be necessary to advise the Americans, Bevin said, while letting them 'think that it was they who were acting'.[15]

On the same evening, Bevin met Marshall and impressed upon him the need for some form of Western democratic system comprising America, Britain, France, Italy and the Dominions. This would not be a formal alliance, but an understanding backed 'by power, money and resolute action'.[16] It would be a sort of spiritual federation of the West. Marshall agreed in general with Bevin's views, but wanted more specific proposals before committing the United States to any formal step. He therefore sent John Hickerson, the Director of European Affairs in the State Department, to the Foreign Office the next day to try to find out more about Bevin's ideas. Hickerson was a sharp-minded and incisive Texan who had long experience in dealing with Western Europe. He was told that Bevin envisaged 'two security arrangements, one a small tight circle including a treaty engagement between the UK, the Benelux countries and France. Surrounding that, a larger circle with somewhat lesser commitments in treaty form bringing in the US and Canada also.'[17] This was the Atlantic pact idea.

Despite earlier public expressions of support for some form of Western collective security pact, it is difficult to disagree with Sir Nicholas Henderson's contention that the broad ideas which Bevin imparted to Bidault and Marshall on 17 December were 'the starting point for the discussions on the security of the Western world which were to begin soon and to last for the following year and a half'. It is also true, however, that despite the elucidation gained by Hickerson in his visit to the Foreign Office on 18 December, Bevin's views at this stage were still somewhat vague and ill-defined. Even within the Foreign Office itself there was a certain amount of confusion about what was meant by 'a spiritual federation'.[18] Also little attempt had been made to establish the priority between the European and Atlantic strands of policy.

The Foreign Secretary attempted to clarify his ideas a little in a series of papers placed before the Cabinet on 8 January 1948 and in a memorandum sent to the British ambassadors in Paris and Washington on 13 January.[19] The Soviet Union, he said, had formed a solid political and economic bloc and there was no prospect in the immediate future of reestablishing and maintaining normal relations between the countries either side of the Soviet line. What was needed was some form of union in Western Europe, either of a formal or an informal character, backed by the Americans and the Dominions: 'in Europe this union should comprise, in addition to the UK: France, the Benelux countries, Italy, Greece, the Scandinavian countries and possibly Portugal. As soon as circumstances permitted, the British Government would also wish to include Spain and Germany, without which no Western system would be complete.'[20]

It is clear from these Cabinet papers in early January 1948 that the Foreign Secretary continued to seek the creation of a 'Third Force' which would balance the roles played by the United States and the Soviet Union. In an important memorandum on 4 January entitled 'The First Aim of British Foreign Policy' Bevin emphasised that although material assistance from the United States was crucial, 'the countries of Western Europe . . . despise the spiritual values of America'. The Foreign Secretary went on to argue that:

> Provided we can organise a Western European system . . . backed by the resources of the Commonwealth and the Americas, it should be possible to develop our own power and influence to equal that of the United States of America and the USSR. We have the material resources in the Colonial Empire, if we develop them, and by giving a spiritual lead now we should be able to carry out our task in a way which will show clearly that we are not subservient to the United States or the Soviet Union.[21]

Bevin's note to the Americans later in January developing his ideas is particularly revealing. Much of the 4 January memorandum was reproduced but the Foreign Secretary deliberately omitted the section (quoted above) dealing with the search for a role independent of the United States. Instead Bevin called for the creation of a Western Union backed by 'power, money and resolution' and 'bound together by the common ideals for which the Western Powers have twice in one generation shed their blood'.[22]

In a practical sense the Foreign Secretary continued to believe that the first step towards this union should involve Britain and France signing separate bilateral treaties with Belgium, Holland and Luxembourg on the lines of the Dunkirk model. On 16 January it was decided to approach the Benelux countries with this proposal.[23] Bevin believed, however, that the coordination of Western Europe represented the nucleus of a much broader arrangement which would involve cooperation between European countries, the Middle East and the development of African resources.

On 22 January the Foreign Secretary unveiled his plans in public for the first time to the House of Commons.[24] In his speech the Foreign Secretary spelled out to the House more plainly and bluntly than ever before the nature of Soviet tactics to extinguish opposition in Eastern Europe and the techniques of political warfare which were being used to intimidate states around the periphery of the Soviet sphere of influence. It was for this reason, he argued, that the 'free nations of Western Europe must now draw together'.[25] Britain could not stand outside Europe and regard its own problems as quite separate from those of its European neighbours. It was important for Britain and France to take the initiative to rebuild a healthy and self-reliant Europe in order to avoid 'the devils of poverty and disease' which would create the conditions for 'war and dictatorship'.[26] In a reference to Soviet obstruction, especially over the Marshall Plan, Bevin struck a defiant tone in the final words of his speech. He argued that 'we shall not be diverted, by threats, propaganda or fifth-column methods, from our aim of uniting by trade, social, cultural and all other contacts those nations of Europe and the world who are ready and able to co-operate. The speed of our recovery and the success of our achievements will be the answer to all attempts to divide the peoples of the world into hostile camps.'[27]

The speech was well-received both in Britain and the United States. At home, *The Economist* talked of the 'boldness, hope and determination' which characterised Mr Bevin's remarks to the House.[28] The Foreign Secretary, it was argued, had 'set the faltering pulse of Western Europe beating more strongly . . . [and] headed British foreign policy in a new direction'.[29]

In the United States the Foreign Secretary's initiative was also 'warmly applauded'.[30] Marshall replied that he was not only very interested in the proposal but 'deeply moved' by it and promised to see that the United States would do everything it properly could 'in assisting the European nations in bringing along this line to fruition'. In a conversation between Marshall and Lord Inverchapel, the British ambassador in the United States, on 19 January, the American Secretary of State said that he was 'turning over in his mind what procedure was likely to be most effective and at what point he could suggest the participation of the United States government in the plan'.[31]

In practice, however, it soon became clear to Bevin that the American position was somewhat more cautious than Marshall's comments suggested. Within the State Department, Marshall himself, together with Under-Secretary Robert Lovett, Charles Bohlen, the Counsellor in the State Department, and George Kennan, the Director of the Policy Planning Staff, all regarded the idea of a military alliance with Western Europe as premature. Even John Hickerson, who was strongly in favour of a military alliance, was not in favour of Bevin's plan to conclude treaties based on the Dunkirk model with the Benelux countries. Such treaties, he argued, directed as they were solely against Germany, were 'unreal and inadequate'.[32] They would be likely to upset Congressmen in the United States who were opposed to 'the real threat to Western Europe which came from further East'.[33] What was needed, Hickerson told Inverchapel, was a multilateral defence pact among West European states based on the model of the Inter-American Treaty of Rio de Janeiro.

According to Hickerson, 'the most important thing was that any American contribution should be made only at the clearly expressed wish of the Europeans themselves . . . Every proof that the free states of Western Europe could give that they were resolved and able to stand on their own feet' would be more likely to secure American assistance.[34] The implication was clear. Even from the most ardent supporter of Bevin's initiative in the administration, the message was that Britain and the West European states would have to organise themselves before any US assistance could be contemplated; and the preferred model was the Rio Treaty, not the Dunkirk Treaty.

For Bevin, this response was not only disappointing, it created a fundamental dilemma. He knew full well the domestic factors in the United States which lay behind the reluctance to make a formal commitment to Western Europe. Traditional reluctance to enter 'entangling alliances' was reinforced by immediate political considerations, with the European Recovery Program not yet through Congress and with the growing shadow of the Presidential elections. Bevin realised that it was only natural that the US government should evince a certain caution in its approach to 'so historically contentious a question as military commitments on the continent'.[35] At the same time, however, he believed that some form of American commitment was essential for the success of his proposals. In a reply to Inverchapel on 5 February, he explained the dilemma clearly.[36] There was a risk of getting into a vicious circle, he explained: without the assurance of security, which could only be given with some degree of American participation. Britain was unlikely to be successful in making the Western Union a going concern. But until such success transpired, the Americans would not discuss participation.

The Foreign Secretary's determination to secure US participation in his plan continued throughout February and early March. If anything, however, the American support which had been so clearly expressed in January seemed to the British to be waning during this period. In a meeting between Inverchapel and Lovett on 2 February, the American Under-Secretary of State reiterated the view that *when* there was evidence of unity with a firm determination to effect an arrangement under which the various European countries were prepared to act in concert to defend themselves, the United States would 'carefully consider the part it might appropriately play in *support* of such a Western European union'.[37] Previously the term 'participation in a Western Union' had been used. Now it was 'support of' a Western Union. This was seen in Britain as indicating that the Americans were now thinking in terms of 'ultimately reinforcing rather than of taking a direct share in whatever European defence arrangements may emerge from the 'initiative'.[38] The Americans clearly understood the impact which their backtracking had on the British. On 7 February Lovett went out of his way to explain American domestic difficulties to Inverchapel and to ask the ambassador to tell Bevin that Britain should not believe that they were half-hearted or 'coy' about his proposals.[39] From Bevin's point of view, however, the question of what concrete support Britain would get from the United States was crucial.[40]

At much the same time as the Foreign Secretary was facing difficulties with the United States, he was also under pressure from the Benelux countries. Apart from their desire to know what American and British help would be forthcoming in the event of war, they, like the Americans, were not keen on the Anglo-French proposals for bilateral treaties on the Dunkirk model. The Benelux countries welcomed Bevin's speech in Parliament of 22 January and expressed their desire to respond as 'sympathetically and as helpfully as possible'.[41] According to M. Spaak, the Belgian Prime Minister, it was 'an epoch-making event of major political and historical importance'.[42] The Benelux countries, however, were clear that the Dunkirk model was inappropriate. That treaty was directed against Germany and as such, in their view, did not correspond to the current realities. Echoing American views, what was required, according to the Benelux countries, was a regional organisation of Western Europe under Article 52 of the United Nations Charter. The Rio Treaty seemed to be a more appropriate model to them, as well as to the Americans.

In early February 1948, however, the British remained highly critical of the notion of a regional pact.[43] The Foreign Office had four main objections. First, that it was unreasonable to suppose that a regional agreement confined to Western Europe could assure security against Russia. Secondly, continuing their wartime concern, that it was 'unnecessarily provocative to the Russians'. Thirdly, that a Western Union had to be built up in stages and that a regional pact meant moving too fast. And finally, that it might encourage American isolation: 'We do not wish by concluding a regional pact of this character to encourage the American school of thought which believes that Western security can be sufficiently assured by a Western regional pact without American participation.'[44] This was a concern that Bevin himself expressed on a number of occasions in the early part of February.[45] The Foreign Secretary was aware, however, that pressure from both the United States and the Benelux countries on this matter could not be ignored. As a result, from 13 February onwards more and more attention was given in the Foreign Office to the Rio Pact and other possible collective pacts under Articles 51 and 52 of the UN Charter.[46]

A number of problems persisted, however, to prevent wholehearted acceptance by Britain of the regional pact plan. The main difficulty was the attitude of France.[47] The French themselves remained committed to the Dunkirk model and were very sensitive over the question of Germany. A regional pact directed specifically against the Soviet Union, which did not deal with the possibility of a future rearmed Germany, was not one which the French could accept. Neither was concern about Germany confined to French circles. The Foreign Secretary himself was concerned to have a treaty designed as 'a protection against possible German aggression, since this would avoid any risk of the appearance of the countries of Western Europe "ganging up" against Russia'.[48] Even at this stage of the Cold War, despite what revisionist historians have argued, Bevin still considered it necessary to 'leave the door open to a Russian conciliation with the West'.[49]

As a result of discussions with the French it was agreed on 16 February that draft treaties on the Dunkirk model should remain the basis of the Anglo-French negotiating position with the Benelux countries. This, however, was seen in Britain as providing a 'basis of discussion' and not an immutable position.[50] By this time it was recognised in the Foreign Office that, on balance, the Treaty of Dunkirk was probably not the best instrument to achieve the Foreign Secretary's objectives.[51] The change in position was largely the result of a growing conviction that a multilateral treaty was much more likely to encourage American participation and prevent a return to isolationism. It would seem also that the Czechoslovakian *coup* in late February influenced the French in favour of a wider formula for West European security than that provided by the Dunkirk Treaty.[52]

The impact of the Czechoslovakian crisis on concentrating the minds of Western leaders at this time cannot be exaggerated. In less than a hundred hours the whole character of the Czech state was transformed. For Bevin the major concern was the effect that those events would have on France and Italy, where morale was deteriorating rapidly. He believed that Western Europe was 'now in the critical period of 6–8 weeks which . . . would decide the future of Europe'.[53] Moreover, Czechoslovakia was not the only problem facing the Western leaders. Concern was also growing in early March about Soviet actions in Berlin; there was fear of a Prague-style *coup d'état* in Italy; and both Finland and Norway faced Soviet pressure to sign treaties of friendship and mutual assistance with the Soviet Union, not unlike those signed by Hungary and Romania. As Lord Bullock has aptly remarked:

> We do not know, and perhaps never shall, whether these different Soviet moves formed part of a concerted plan (probably not) or what were the Soviet objectives in making them and following them with the blockade of Berlin. But it was impossible for Bevin and other men in office in Western Europe and the USA not to put the pieces together and act on the assumption that they were facing a calculated challenge by the Russians.[54]

Bevin himself did not believe that the Russians wanted war, but he believed that war was nevertheless a distinct possibility unless the West acted firmly. Resolute action was necessary to prevent the slide from crisis to crisis which had occurred in the 1930s. The Foreign Secretary was anxious lest the period of early March 1948 would be 'the last chance for saving the West'.[55]

The effect of these dramatic and dangerous events was to secure acceptance by both Britain and France that a multilateral pact was the way forward for Western Europe. This is not to say that there were no disagreements among the five powers when negotiations began in earnest on 4 March. Two problems in particular preoccupied the negotiators through to 12 March, when the draft Five Power Treaty was agreed. The first concerned attitudes to Germany; the second concerned the future development of the pact.

On Germany, the Benelux countries went out of their way to reject the kind of references to that country included in the Dunkirk Treaty.[56] They urged instead the need to make reference in the new treaty to the possibility of a regenerated Germany entering the European comity of nations. The French, on the other hand, were very much opposed to any reference to the possibility of eventual collaboration with a regenerated Germany and urged that the 'menace of German aggression' should be included in the treaty.

The British position was that the German danger could not be ignored and that past German aggression had to be mentioned (as it was, in Article 7). At the same time, Bevin supported the reference to possible German regeneration and its return to the European system of states.[57]

The second disagreement concerned the future expansion of the pact.[58] The Benelux countries saw the pact primarily as a means of tightening up the relations between the five powers. They were not hostile to the idea of it forming the nucleus of a wider organisation of Western Europe, but they were not, at this stage, ready to regard other powers 'such as Italy or Portugal, as sufficiently closely allied to them by interest or tradition to participate automatically in the arrangements they would welcome between themselves on the one hand and the UK and France on the other'.[59] Bevin's position, however, was that the pact should be seen as being 'very much wider in scope than a defensive military agreement between the Five Powers'.[60] He saw the treaty as being the basis for the organisation of Western Europe as a whole, which could be expanded to include other West European states. He also saw it more in terms of a convenient device to convince the Americans that West European states were prepared to stand on their own feet. That, he hoped, like his initiative after Marshall's Harvard speech, would encourage direct American participation.

Bevin's attitude towards the pact is seen very clearly from his reaction to the crisis generated over Soviet pressure on Norway to sign a defence pact. On 11 March 1948 he suggested to the Americans and Canadians that the new round of Soviet demands indicated that time was running out if the Soviet Union was to be contained.[61] What was needed was a Regional Atlantic Approaches Pact of Mutual Assistance (under Article 51 of the UN Charter) which would include all those countries threatened by a Russian thrust to the Atlantic – the United States, the United Kingdom, France, Canada, Ireland, Iceland, Norway, Denmark, Portugal and Spain (when it embraced democratic government). For Bevin, the Brussels pact was not enough on its own.

By 12 March the minor differences facing the delegates to the Five Power conference were resolved, and on 17 March Bevin travelled to Brussels via Paris with M. Bidault to sign the Brussels Treaty. For the British Foreign Secretary the Treaty was an important achievement, but, as he told Bidault, there was still an urgent need to get the United States to underwrite the pact: 'The object should be to ensure that if there was a war, the US should be in from the first day and that we should not have to wait another Pearl Harbour . . . we must bring the US to face up to their responsibilities.'[62] The initial reaction in the United States to the signing of the pact was therefore comforting to the British. President Truman told a special meeting of Congress on 17 March that the developments in Europe deserved the full support of the United States. 'I am confident', he said, 'that the US will, by appropriate means, extend to the free nations, the support which the situation requires. I am sure that the determination of the free countries of Europe to protect themselves will be matched by an equal determination on our part to help them do so.'[63]

In any assessment of the significance of the Brussels Pact an important question arises. Was the Five Power Treaty 'a sprat to catch a whale'? Was it simply a useful device used by Bevin for the sole purpose of involving the United States in the defence of Britain and Western Europe? Or did the Foreign Secretary envisage other roles for the 'Western Union'?

With hindsight it is easy to come to the conclusion that the Brussels Pact had a single purpose: to demonstrate to the United States, following Hickerson's comments, that the Western European states were serious about their own defence and determined to try to stand on their own feet. This interpretation of the Brussels Pact is favoured by Elizabeth Barker in *Britain between the Superpowers*.[64] In her view the Western Union was designed above all else 'to lure the Americans into giving Western Europe full military backing in the face of the Soviet threat, in much the same way as the joint West European response to the Marshall offer had procured US economic aid.'[65]

There is clearly some validity in this view but it tends to ignore other important purposes which the Brussels Pact was designed to promote. Even though the Foreign Secretary was clearly intent on drawing the United States into the defence of Western Europe, from late 1947 he continued to be tempted by the idea of Britain playing a role somewhat independently of both the United States and the Soviet Union. He also remained interested in pursuing British leadership in Western Europe.

For both economic and security reasons Bevin and his colleagues recognised that American support was essential if the Western Union idea was to be successful. Nevertheless, the Cabinet distinguished between short- and long-term objectives. On 5 March 1948 it was argued that while 'we shall use the United States aid to gain time . . . our ultimate aim should be to attain a position in which the countries of Western Europe could be independent both of the United States and the Soviet Union'.[66] Geoffrey Warner has argued that 'although this remark cannot be attributed specifically to Bevin, the sentiments it contained certainly seem to underlie much of his thinking'.[67]

On 22 September 1947 Bevin told the French Prime Minister, M. Ramadier, that if the European colonial powers, especially Britain and France, pooled their African resources they could be 'as powerful as either the Soviet Union or the United States'.[68] A few months later in a conversation with M. Bidault, the French Foreign Minister, Bevin 'drew attention to the great resources of Western Europe both in Europe and their African Colonies. If properly developed these resources amounted to more than either the Soviet Union or the United States could muster, and should enable Western European Powers to be independent of either.'[69]

It is also clear that Bevin had a genuine interest in using the Brussels Pact as a vehicle for his ideas about greater European cooperation. Although he was forced to abandon his plan for a customs union because of opposition from Cabinet colleagues he nevertheless continued to press for the closest possible financial integration among the five Brussels Pact powers.[70] On 27 February 1948 he told the new Chancellor of the Exchequer, Sir Stafford Cripps, that it was necessary to take steps 'to prevent the various potential partners in Western Union . . . from following completely different economic policies'.[71] He followed this up on 17 April, by telling the French President, Vincent Auriol, 'that he believed in the possibility of a Western Union bank and currency and of other economic arrangements designed to make the group more independent of the United States'.[72]

Bevin's rather pragmatic view of European cooperation, which he hoped would emerge as a result of a careful process of inter-governmental negotiations, clashed, however, with the more grandiose visions of the European federalists. The Foreign Secretary had little time for blueprints, written constitutions and 'neat-looking plans on paper' designed to achieve Western European unity.[73] He told the House of Commons on 15 September 1948 that

> the intricacies of Western Europe are such that we had better proceed . . . on the same principle of association of nations that we have in the Commonwealth . . .. I think that adopting the principle of an unwritten constitution, and the process of constant association step by step, by treaty and agreement and by taking on certain things collectively instead of by ourselves, is the right way to approach this Western Union problem. When we have settled the matter of defence, economic co-operation and the necessary political developments which must follow, it may be possible, and I think it will be, to establish among us some kind of assembly to deal with the practical things we have accomplished as Governments, but I do not think it will work if we try to put the roof on before we have built the building.[74]

Faced with growing pressure from the federalists as a result of the Hague Congress and a French proposal for a European parliamentary assembly,

Bevin eventually put forward an idea of his own in September 1948 for a Council of Western Europe. This would consist of an annual meeting of Ministers of the Brussels Treaty Powers which would 'have in general a power of initiation, and should be in a position to recommend measures for acceptance by the group as a whole'.[75] He saw the Council as a 'cabinet for Western Europe' which would be 'the final storey in the structure set up under the Brussels Treaty'.[76] When Bevin's proposal was discussed by the Brussels Treaty Powers in late 1948, however, a rather messy compromise was agreed between the French proposal for a European assembly and Bevin's Council of Western Europe. A Council of Europe emerged with a Council of Ministers and a parliamentary assembly with delegates rather than directly elected members.[77]

What this suggests is that the Brussels Pact was both an attempt to encourage American involvement in European security and simultaneously part of the British government's pursuit of a leadership role in Western Europe (which, in turn, was perceived as a core element of the Foreign Secretary's plan for a broader Western Union). These were not seen as contradictory objectives. The government clearly retained a desire to pursue an independent role, especially in the economic field. In the short term close relations with the United States were necessary to create the conditions for British leadership of Western Europe and help lay the foundations for a Western Union. There seems to be a great deal of substance in the view, however, that the Brussels Pact was more significant as a step to achieve a Western Union than as a first step to establish an Atlantic Alliance. According to John Kent and John Young, 'it was not that Bevin needed a European alliance in order to win a US alliance. The case was quite the *opposite*: a US alliance was needed in order to make a European-based system, the Western Union, effective. And the Western Union was intended to maintain Britain's standing as a major power, independent of the United States.'[78] Increasingly, however, the perceived threat from the Soviet Union overshadowed and coloured every aspect of Britain's view of Europe. As the international climate darkened, the pursuit of Western European leadership and a more independent role became more difficult. The more immediate concern for security pushed the government towards a longer-term Atlanticism which had been consistently advocated by the Chiefs of Staff since 1944. The continuing preoccupation with security was to ensure that it was this Atlanticist option which eventually prevailed rather than the European option. In order to understand the growing emphasis given to Atlantic security arrangements it is necessary to consider the evolution in the thinking of the CoS about European security in the postwar period and the continuing debates which took place between the military and the diplomats.

6 The Chiefs of Staff and the Continental Commitment

In the period from 1945 to 1948, defence planners in Britain struggled to develop a coherent defence policy in a period of great uncertainty and transition. The key debates of the time centred on five interrelated issues. These included the conflict between defence and economic reconstruction, Commonwealth defence, Middle Eastern strategy, relations with the United States and the question of a continental commitment. On these issues the disputes which occurred sometimes involved clear differences between the Foreign Office and the Chiefs of Staff but on other occasions cut across military and civilian departmental boundaries. Differences emerged within the Cabinet, and the Chiefs of Staff Committee over the way forward for British defence policy. In order to understand the position adopted by the CoS on European defence it is important to look at each of the issues in turn and the differences which occurred.

After the devastation of the war, one of the central problems facing Britain in the immediate postwar period was how to rebuild the economy while continuing the numerous defence responsibilities which remained as a legacy of the conflict. The war had cost Britain a quarter of her national wealth. It had been necessary to sell foreign assets worth £4200 million in order to pay for supplies bought from abroad during the war. Her debts to other countries had increased from £476 million in August 1939 to £3355 million in June 1945. To all intents and purposes Britain was insolvent for the first time in her history. Yet as one of the victorious powers in an international environment which remained very uncertain she felt obliged to make a major military contribution to the tasks of rebuilding a new, more stable international order. In February 1946 Attlee highlighted the direct conflict of requirements facing Britain when he told the Cabinet that 'we cannot abandon our responsibilities in many parts of the world: to do so would be to throw away the fruits of victory and to betray those who fought in the common cause'.[1] At the same time, however, he stressed that the need for economic rehabilitation was vital. This meant that demobilisation had to be speeded up to find the manpower so desperately needed by British industry.[2]

Within the Cabinet this dilemma reflected the differing responsibilities of the Treasury and the Board of Trade on the one hand and the Foreign Office and Ministry of Defence on the other. In January 1946 Hugh Dalton, the Chancellor of the Exchequer, told the Defence Committee that Britain was nearly one million men short of the minimum required to achieve the revival of Britain's export trade. Quite apart from the problem of manpower he also emphasised the enormous burden incurred on the balance of payments through overseas expenditure on defence. Dalton argued that if there was not a quick drop in defence spending he 'could not promise anything but economic disaster'.[3] These were warnings he repeated constantly throughout 1947 as the scale of Britain's economic difficulties grew greater.

These pressures to cut defence expenditure and service manpower were resisted by Bevin and the Defence Minister, A. V. Alexander. The Foreign Secretary argued that until peace treaties were concluded (and this task was proving extremely difficult because of Soviet intransigence) British troops could not be withdrawn from their occupational roles. Major reductions would 'seriously undermine his position in discussions with the Russians' and weaken the military backing so important for the conduct of an effective foreign policy.[4] By November 1946 Bevin and Alexander were arguing that Soviet expansion and the growing gaps between commitments and capabilities were such that compulsory national service was necessary.[5] This measure, unprecedented in peacetime in Britain, was accepted even though it was recognised that it would inevitably hamper economic recovery and the tasks of reconstruction.

Faced with these difficulties in the early postwar period the government and the CoS searched for ways to lessen the burden of Britain's global defence responsibilities. One idea which had been discussed from 1944 onwards was Commonwealth cooperation. The initiative for emphasising the importance of the Commonwealth, as a central pillar for restructuring Britain's world power after the war ended, came from Attlee and Bevin. In a Cabinet paper in June 1943 Attlee argued that it was 'a fundamental assumption that whatever post-war international organisation is established, it will be our aim to maintain the British Commonwealth as an international entity, recognised as such by foreign countries'. He went on to argue that if Britain was to carry its full weight in the postwar world with the United States and the USSR it could 'only be done as a united British Commonwealth'.[6] Much the same line of argument was developed by Bevin, as Minister of Labour, during 1944. When the issue was discussed by the Vice-Chiefs of Staff on 6 March 1944, Ian Jacob of the War Cabinet Secretariat summarised the Minister of Labour's arguments in the following way:

> Mr. Bevin points to self-preservation as being the first and strongest instinct, and suggests that if the British Empire could act as a unit for

> self-preservation purposes, trade, finance, and constitutional developments would flow from it. He considers that a system of regions or zones, which would enhance the sense of responsibility of the Dominions, would tend to facilitate the driving together of the Empire. In the earlier part of his Paper, it was suggested that through obligations under a world system of security the desirability of close consultation and co-operation in defence matters would be brought home to those Dominions who would otherwise be reluctant to give up any part of their independence. Mr. Bevin's idea might thus eventually be achieved by indirect means.[7]

It was this initiative which led the Joint Planning Staff to produce a paper on 'The Co-ordination of Defence Policy within the British Commonwealth in Relation to a World System of Security'. The matter was also raised in the May 1944 Prime Ministers Commonwealth Conference. Due to the opposition of Canada in particular, however, the proposals were not pursued further during the war.

According to John Albert, it was Attlee who was responsible for the revival of the idea of the Commonwealth 'as the core of a world security system' after the war ended.[8] For the new Prime Minister the Commonwealth was an ideal model for enhancing international order in the post-war world. On his initiative, the CoS revived the idea of Commonwealth defence cooperation in October 1945 and recommended that an approach should be made to Canada, South Africa, Australia and New Zealand with a view to coordinating national defence policies. This was followed up in March 1946 when the Service Chiefs adopted General Sir Alan Brooke's suggestion that a report should be written by Joint Planners based on the idea of 'zones of defence' in the Middle East, India, South East Asia and Australia, coordinated from London.[9] This was produced on 27 March and approved by the Defence Committee as the basis for discussions at the Commonwealth Prime Ministers Conference in April and May.[10]

The aim, however, of getting the Commonwealth countries to share the defence burden and so help Britain to continue to play the role of a Great Power, in the common interest, once again proved difficult to achieve. Despite some positive response from Australia, and particularly from New Zealand, the sensitive problems of sovereignty associated with a common defence policy continued to be opposed by the Canadians. Julian Lewis has argued that the discussions at the London meeting of Commonwealth leaders in April and May 1946 showed that 'there was little prospect at this time of concluding definite political agreements with members of the Commonwealth which would bind them together when its security was endangered'.[11] The notion of a Commonwealth global security policy, however, was to become central to British strategic planning.

In June 1947 the idea of a Commonwealth defence policy was officially accepted with the CoS endorsement of 'The Overall Strategic Plan'.[12] According to this major defence document, which was to remain the bedrock of British strategic planning for the next three years, a Commonwealth defence policy centred on three main pillars: the defence of the United Kingdom, the control of essential sea communications and a firm hold on the Middle East. In this global security framework a united Commonwealth and greater defence cooperation between Commonwealth countries were of great importance.

In line with this another attempt was made to coordinate the defence efforts of the Dominions at the October 1948 meeting of the Commonwealth Prime Ministers in London. The British Chiefs of Staff told the Dominion leaders that if war broke out with the Soviet Union as a result of the Berlin 'blockade' five main aims should be pursued in cooperation with the allies. These were: 'to secure the integrity of the Commonwealth countries; to mount a strategic air offensive; to hold the enemy as far east as possible in Western Europe; to maintain a firm hold on the Middle East; and to control essential sea communications'. The CoS argued that this required the preparation of common strategic objectives and the coordination of military plans in peacetime if allied cooperation was to be effective in war.[13] This time British policy was more successful.

At the end of the Conference a consensus had been achieved: that greater cooperation between Britain and the Dominions was necessary.[14] In the months that followed, a peacetime system of defence liaison between the members of the Commonwealth along the lines envisaged by the British from 1944 onwards began to emerge.[15] The process of cooperation, however, remained extremely slow. By 1949, despite some limited progress, the structure of Commonwealth defence still remained 'rudimentary'. The high hopes of British planners had not yet been fully achieved.

At the time that efforts were being made to coordinate the defence policies of the Dominions the major strategic preoccupation of the CoS and the government was the Middle East. In early 1946, Soviet pressure on Turkey and Persia, and demands for control over Tripolitania, helped to focus the attention of defence planners on the Middle East and a Mediterranean strategy. In line with the wartime planning the CoS saw the Middle East as second only to Britain itself in terms of overall defence policy. Every effort therefore had to be made to prevent Soviet penetration in the area. It was regarded as an area of vital strategic interest to Britain for three main reasons. Firstly, because of its geographical position in relation to the

Commonwealth; secondly, because of the growing importance of its oil both in peacetime and in war; and thirdly because of its potential as a base for a strategic air offensive against the Soviet Union should war break out.[16] This stress on the importance of the Middle East as a major base to attack the Soviet Union in the event of war was highlighted by the CoS in July 1946. They argued that:

> If Russia were the enemy, her great superiority of manpower would necessitate our making the maximum use of our technical and scientific superiority and particularly of our air power to strike back at the enemy's vital centres, war potential and communications. In a war with Russia, bases in the United Kingdom alone would not be sufficient for this task. The Middle East provides the only air bases from which effective offensive action can be undertaken against the important Russian industrial and oil-producing areas of Southern Russia and the Caucasus.[17]

Disagreement, however, arose in February 1946 when Attlee sent a paper to the Chiefs of Staff arguing that their views were outdated. The Prime Minister regarded the emphasis given by the CoS to the Middle East as a legacy of the old imperial strategy, which no longer suited Britain's interests. He was not convinced that Mediterranean communications were a vital British interest, or that they could be safeguarded, with the advent of air power. Such a strategy was also too expensive for an impoverished country like Britain. 'I consider that we cannot afford to provide the great sums of money for the large forces involved on the chance of being able to use the Mediterranean route in time of war', Attlee argued. 'We must not for sentimental reasons based on the past give hostages to fortune. It may be we shall have to consider the British Isles as an easterly extension of a strategic area, the centre of which is the American continent, rather than as a power looking eastwards through the Mediterranean to India and the East.'[18]

This 'heroic' attempt to scale down British imperial strategy was vehemently opposed by the CoS and by the Foreign Secretary.[19] In their reply in late March the CoS argued that it would be a major error 'to cut our commitments and thereby lose our predominant position in such areas'.[20] If we do this, important strategic bases would soon be lost to the Soviet Union. This would result in Soviet domination of all of Europe (apart from Britain), North Africa as well as the Middle East. Britain would then be faced with a 'threat to its sea communications, coupled with the direct threat by air attack and long-range bombardment from the mainland of Europe . . . the United Kingdom would be reduced to a Malta-type existence, contributing little to the main war potential'.[21] According to Field Marshal Montgomery the CoS felt so strongly about the Prime Minister's challenge to the primacy of the Middle Eastern strategy that they threatened to resign over the issue. Whether this happened is rather doubtful but there is no denying the determination of the CoS to stand their ground over the importance of the Middle East in British strategic policy.[22]

The Foreign Secretary supported the CoS against the Prime Minister on the issue. Bevin was worried by the prospects of greater Soviet involvement in the Middle East and the serious political affects which would result from Britain's withdrawal from the region. He felt strongly that 'our presence in the Mediterranean serves a purpose which is vital to our position as a Great Power. The Mediterranean is the area through which we bring influence to bear on southern Europe, the soft under-belly of France, Italy, Yugoslavia, Greece and Turkey. Without our physical presence in the Mediterranean, we should cut little ice in those states which would fall, like eastern Europe, under the totalitarian yoke. We should also lose our position in the Middle East (including Iraqi oil, now one of our greatest assets), even if we could afford to let Egypt go. If we move out, Russia will move in.'[23]

Faced with the combined opposition of the CoS and the Foreign Secretary, the Prime Minister yielded 'gracefully'.[24] The CoS paper which was prepared for the Commonwealth Prime Ministers Conference in April and May once again emphasised the importance of a British presence in the Middle East for both political and military reasons. The paper argued that it helped to prevent the expansion of Soviet influence in the area. From a military point of view, bases in the Middle East were also very useful to launch long-range attacks not only on the 'important industrial and oil producing areas of southern Russia and the Caucasus, but also many other important industrial areas of Russia'.[25] The central importance of the Middle East was also highlighted in 'The Overall Strategic Plan' in June 1947 which was to remain the guiding defence document until 1950.

It seems highly likely that this emphasis on the military value of Middle East bases was, in part, conditioned by the CoS hope that the United States would fight alongside Britain in a future conflict. In any atomic air offensive against Russia, Middle Eastern bases would be of great value to the United States. Despite the cooling of the political relationship between the two countries after the war ended the CoS retained a firm conviction that close military ties were of vital importance to British security interests. As Elizabeth Barker has argued 'the need to secure American co-operation in the Middle East, the Pacific, Europe and the atomic field was a driving force in British strategy as far as the Chiefs of Staff' were concerned.[26]

In mid-1947 a close military alliance between Britain and the United States was little more than an article of faith. When Winston Churchill called for a special defence relationship to be established between the two countries in his Fulton Speech in March 1946 the scale of military cooperation which existed was very limited. The Combined Chiefs of Staff Committee remained in existence and a range of informal military contacts were developed following Montgomery's visit to the United States in September 1946 but there was little political support in either country for a broader political alliance. Certainly there was no commitment by the United States to come to Britain's assistance if war broke out.[27] Bevin supported the development of military ties between the two countries but remained uncertain about the future direction of American foreign policy.[28] It is remarkable therefore that the CoS continued to base their strategic planning from 1946 onwards on the assumption that the United States would be an ally in any future war.[29] In January 1947 the Service Chiefs argued that the moment of American entry into a future war on Britain's side could not be forecast but as in previous conflicts in the twentieth century, in their view, US assistance would finally be forthcoming. The aim was to make sure that the period when Britain had to act alone was kept to the absolute minimum.[30] By July all three Services were planning on the basis of 'quite considerable assistance from the United States'. The CoS regarded this as a 'reasonable' basis for planning because 'in our view, the US would never allow the UK to be isolated, even if they did not enter the war at the same time as us'.[31]

The first concrete signs of greater Anglo-American security cooperation came in October 1947 with the Pentagon Talks on the Middle East. At these talks the British fully succeeded in convincing the United States of the importance of the Middle East and in so doing opened the door for a greater degree of strategic planning on a global basis between the two countries.[32] By April 1948 Britain and the United States were working on parallel emergency war plans. For the British Chiefs of Staff, 'Commonwealth' defence implied not only the attempt to achieve the strategic unity of the Commonwealth but also a close alliance with the United States.

What role, however, was Western Europe to play in this broad concept of Commonwealth defence? In 'The Overall Strategic Plan' in June 1947 the military chiefs spelled out their attitude towards a Western European group.[33] As in 1944 they supported such a group because it would 'delay the enemy's advance across Europe'. Britain, they argued, should 'encourage the building up of a strong Western Region of Defence with France as its keystone, and ensure that Germany does not become a Russian satellite'. Their reservations, however, remained that a regional organisation of this kind would not be capable of matching the tremendous power of the Soviet Union. 'No European bloc could, as of old, be relied upon to resist aggression effectively'. In their view in any future war active and early support from the United States would be indispensable. 'American manpower, industrial resources and supplies of mass destruction weapons alone could turn the scale' against the most powerful state on the continent.[34] Despite the value of a Western European bloc, the CoS stressed the vital nature of Anglo-American military cooperation, especially in the Middle East in any future conflict. The Middle East, not Western Europe, was the key focus of British strategic planning.[35]

Despite the unanimity amongst the Chiefs of Staff on the importance of close military ties with the United States and the inadequacy of a Western European group to provide protection against the Soviet Union, important differences nevertheless emerged over Britain's contribution to European defence. In July 1946 the new CIGS, Bernard Montgomery, had urged his colleagues that support should be given to the formation of a strong western bloc, with the acceptance of a commitment 'to fight on the mainland of Europe, alongside our Allies, with all that that entailed'.[36] The other two Chiefs of Staff, Air Chief Marshal Lord Tedder and Admiral of the Fleet Sir John Cunningham, however, were not prepared to consider this argument. Because the CIGS, General Montgomery, was the newcomer, he felt unable to force the issue.

The question was raised again, however, in December 1946 by Lieutenant-General Sir Hastings Ismay, the Chief Staff Officer to the Minister of Defence. Ismay urged the CoS to consider the politico-strategic significance of a British continental commitment. He pointed out that 'if it was decided that our forces should never again be sent to the aid of a Power on the Continent, France might turn to an Eastern alliance for her future security . . .. The military implications of our relationship with France were, therefore, either to accept that we would in future have to send forces to her aid, or to run the risk that the whole of the European Continent was available to a potential aggressor, and of the latter's ability to attack this country to the point of destruction with air weapons, and to attack our sea lines of communication.'[37]

This was not an idea, however, which commended itself to the Chiefs of Staff, even the CIGS. Despite his earlier support for a continental commitment, in December 1946 Montgomery accepted that British forces 'were too small for operations on the Continent in the early stages of a war'.[38] In his view the 'old concept of operations involving the sending of forces to the Continent was virtually useless against an enemy who possessed almost unlimited man-power'.[39] As far as the other Chiefs of Staff were concerned

British support for the Western European allies would have to be limited to political and economic backing in peace and, in war, the support of naval and air forces.

The prevailing CoS view was outlined in a Report produced by the Future Planning Section (FPS) of the Chiefs of Staff Committee in March 1947. In their view Western Europe was indefensible without American assistance. The United Kingdom therefore should 'avoid being committed to a continental campaign'. Any forces provided for the occupation of Germany should be small and should be part of a strategic reserve. In any emergency they should have to be withdrawn to Britain. Western Europe was therefore to be virtually abandoned if war broke out.[40] This March paper formed the basis of 'The Overall Strategic Plan' DO(47)44 which guided British defence policy from June 1947 to March 1950.[41]

Even though Britain was entering into political commitments with her European allies in 1947 the CoS remained reluctant to commit British forces to the Continent in the event of war. It was mentioned earlier that the Service Chiefs only supported the Dunkirk Treaty provided it did not undermine relations with the United States.[42] Similarly in June 1947, when the Foreign Office asked the CoS if they favoured treaties of alliance with Belgium and Holland on the lines of the treaty with France, the CoS replied that there were advantages in concluding such treaties but 'Britain should not guarantee to send forces to defend them if Russia attacked'.[43] Once again at the end of 1947 when the French urged Britain to open staff talks the CoS 'were decidedly unenthusiastic'.[44] They did not wish to reveal to the French or the Benelux countries that Britain was not planning to stay and fight on the Continent if war broke out.

With Bevin's Western Union initiative in early 1948 the issue reemerged as a major bone of contention between the Foreign Office and the CoS and within the Chiefs of Staff Committee itself. In an important minute written on 9 January, Ivone Kirkpatrick highlighted the concern felt in the Foreign Office:

> The Chiefs of Staff have . . . decided . . . that we shall not in a future crisis initially send a land expeditionary force to the Continent. If this policy is maintained . . . we shall eventually either have to admit to our allies or refuse to disclose our intentions. Both courses are likely to discourage them to the point of refusing to associate themselves with us. In this predicament it seems to me that our only method of satisfying the need for security is to involve America as far as possible in the defence of Western Europe. It is quite likely that the Americans too will refuse specifically to commit their forces to the Continent of Europe; but if they

would enter with ourselves into some general commitment to go to war with the aggressor it is probable that the potential victims might feel reassured as to eventual victory and hence refuse to embark on a fatal policy of appeasement.[45]

For the Foreign Office, at this stage, some form of American commitment to the defence of Western Europe was seen as a useful remedy to the dilemma in which Britain found herself. Agreements were being sought with Western European countries for important political reasons while the CoS were reluctant to provide any military guarantees to the Continent upon which the success of Britain's political initiatives depended.

The question of whether Britain should send an army to the Continent in a future war was raised again in the Defence Committee on 8 January. Bevin argued that it would not be possible to fudge the issue any longer. The moment was ripe for the consolidation of Western Europe and 'he would be embarrassed in his dealings with Britain's potential European allies if planning were based on the assumption that an army was not to be sent' to the Continent.[46]

This was not, however, a view shared by several of the Ministers present. They agreed with the Chief of the Air Staff, Lord Tedder, that the task of maintaining 'an army of continental scale' as well as a first-class navy and air force was beyond Britain's limited resources.[47] In their discussions of 'The Future Size and Shape of the Armed Forces' the Defence Committee decided to proceed on the basis of a very restricted defence budget of £600 million with no provision for a continental army.[48]

Later in the month, the Joint Planners reinforced the views expressed by the Chief of the Air Staff. They pointed out that the Western European states would be eager to get Britain to adopt a continental strategy and make Western Europe 'our main theatre of operations'. In their view, however, this pressure would have to be rejected as militarily impractical 'unless the countries of Western Europe revive both economically and militarily to an extent far greater than appears possible at present'. Their recommendation was unequivocal, that Britain 'should enter into no commitment to send land and air forces to the continent'.[49]

At this stage Montgomery returned to his earlier argument in favour of a continental army. On 30 January he attacked the Joint Planners, arguing that in the event of war the West Europeans would want to 'hold the attack' as far to the east as possible. In this eventuality Britain would need to play its full part and support her allies with the 'fullest possible weight of our land, air and naval power'.[50] The success of the alliance, according to the CIGS, would depend on this assistance.

Montgomery's renewed defence of a continental commitment led to a clash with the other two Chiefs of Staff on 2 February. Tedder and Cunningham both reiterated their obligations to a 'continental strategy' and argued that no decisions could be made until discussions had taken place with the United States.[51] Even after these discussions, they were not optimistic of getting American support for prolonged operations on the Continent. Recent overtures from the US Navy indicated an American interest in joint planning for the evacuation of Anglo-American occupation forces from Western Europe in an emergency. Montgomery replied that even if the Americans did refuse to help in the defence of Western Europe 'we must continue to do our utmost with such allies as we might have'.[52] In his view Britain had a responsibility to take the initiative in these matters rather than slavishly following the American lead.

The same argument continued on 4 February, this time in the presence of Attlee, Bevin and A. V. Alexander. Once again Tedder took the lead, arguing that Britain should maintain an air/sea strategy as its contribution to European defence. He was supported by Cunningham, who emphasised that twice in the past we had given a guarantee to assist a continental nation to the limit of our power by the provision of land forces. On both occasions we had suffered severely, first at Mons and more recently at Dunkirk.[53] Montgomery, on the other hand, in his usual forthright manner, argued that a small British force, of not more than two divisions, would be of major importance to France and the Benelux countries. Such a commitment, he felt, would make a defence of the Rhine a realistic possibility.[54]

Faced with this disagreement between the Chiefs of Staff, the Prime Minister sided with the Naval and Air Chiefs. In his view, 'previous experience had shown how continental commitments, initially small, were apt to grow into very large ones'.[55] Despite his previous reservations about a Mediterranean strategy, he argued that defence of the Middle East would not be possible if land forces were sent to fight on the Continent. Also, two divisions, he felt, would be insufficient to provide any great encouragement to the Western European states.

Bevin was left to find a compromise between the conflicting positions, which would recognise Britain's limited capabilities but also reinforce his Western Union initiative. He argued that the best way forward initially was to find out what forces the West Europeans were able to provide for their defence. These would then have to be organised into 'one effective whole', breaking down the barriers of national pride.[56] At this point if Britain found it had land forces to spare 'he could see no fundamental objections to their fighting on the continent'.[57]

Bevin's view represented something of a fudge on the issue. It proved acceptable, however, to the other members of the Defence Committee and formed the basis of Britain's position during the negotiations which led to the Brussels Pact in March. In the Brussels Treaty Britain pledged to give 'all military and other aid and assistance in its power to the other signatories in the case of an armed attack on any one of them'.[58] At this stage, however, Britain had no plans to send land forces to the Continent in the event of such an attack.

During the negotiations that led to the Brussels Pact, France and the Benelux countries continued to ask what military support Britain would give to the Pact. The Belgian Foreign Minister, Paul-Henri Spaak, in particular, argued that he wanted precise military conversations as part of the treaty negotiations.[59]

The contradictions in British policy were clear to the Cabinet at this time. At a Cabinet meeting which took place on 5 March, shortly after the Communist coup in Czechoslovakia, the Chiefs of Staff were directed to examine their defence priorities in the light of Bevin's foreign policy objectives in Europe.[60] According to one source, 'this crucial meeting . . . launched military planning in a direction not related to Britain's resources (which it never actually had been) nor to an assessment of the Soviet military threat (which in any case was held to be unlikely to emerge) but to the requirements of foreign policy.'[61] The Cabinet was intent on trying to bring defence policy more in line with the nation's foreign policy.

This 'March Directive', however, was only partially successful. On 19 March, two days after the Brussels Treaty was signed, Montgomery persuaded the other two Chiefs of Staff to agree that the aim of the alliance should be to defend Western Europe as far to the east as possible. This meant that in an emergency Britain would fight with the occupation forces it had on the ground.[62] The CoS also agreed that in the event of trouble, it was our intention to remain on the continent.[63]

Despite this 'partial victory' for Montgomery, the Joint Planners returned from a visit to Washington in April, having agreed with their American counterparts to draw up plans for the withdrawal of both British and American forces from the continent in the event of an emergency in the next 18 months. The British plan was code-named Doublequick and the American version was called Halfmoon.[64] For Montgomery, this represented an impossible situation. Britain was about to start planning with its Western European allies for the defence of Europe while simultaneously coordinating plans with the United States to withdraw in an emergency.[65]

Montgomery sought to clarify the situation on 12 May when he asked for a firm decision from the other Chiefs of Staff that Britain should stay and fight on the Rhine if war broke out. He secured the backing of the Defence Minister, Alexander, who agreed that it was not politically acceptable for Britain to sign an agreement with the Western European states and at the same time plan to leave at the first sign of trouble. The CoS therefore accepted that British forces on the continent should remain and fight 'unless and until they were pushed out'. However, in a concession to the Naval and Air Chiefs, who remained sceptical about the commitment, it was decided also that these forces would not be reinforced.[66] It was agreed that Doublequick should be revised to remove the offending references to evacuation and that British forces would stay and fight on the Rhine 'unless and until pushed out' by enemy forces. No reinforcements, however, were to be sent.

This was to remain the British position throughout the Berlin blockade from June 1948 to May 1949 and during the negotiations which led to the North Atlantic Treaty being signed. Despite the pressures from the other Brussels Pact states that Britain should make its position clearer on reinforcements, the arguments continued between the Service Chiefs over the relative priority of the Middle East and Western Europe in Britain's 'Commonwealth defence policy'. Limited resources and pressures from the United States and Western European states increasingly forced Britain to recognise that choices would have to be made within the framework of the overall global strategy.[67] In June 1949 the Chiefs of Staff (minus Montgomery who had become the Chairman of Western Union Commanders-in-Chief Committee) were coming to the conclusion that 'from the broadest strategic point of view it would be right to promise to reinforce the British Army of the Rhine, in spite of the inescapable risk that this will entail'.[68] Tedder, however, still dissented, arguing that the Middle East remained a more important priority and the promise of reinforcements was not necessary to keep the Western Union together.[69]

The first real discernible shift away from the priorities of DO(47)44 can be seen in the meeting of the Chiefs of Staff on 20 April 1949, following the signing of the North Atlantic Treaty. Montgomery's successor, Sir William Slim, proved to be a little more successful in persuading the other military chiefs to accept the need for a greater British contribution to Western European defence. He argued diplomatically that 'although on purely military grounds there was much to be said for keeping the British land forces' contribution to the defence of Western Europe as small as possible in order that we might make the maximum contribution to the defence of the Middle East, there were serious political problems to such a proceeding'.[70] Slim told his colleagues that a choice would have to be made between making a firm commitment to the defence of Western Europe or concentrating on the

Cold War and agreeing to commit only 'left over' resources to strengthen the defence of Europe.

Tedder agreed with Slim that it was no longer possible to avoid the real issue, which was to decide whether or not Britain was prepared to make a major contribution on land to the defence of Western Europe. He warned, however, that any understanding into which Britain might enter to send further forces to the continent would inevitably have repercussions on her ability to carry out other essential tasks in other parts of the world. He argued that the Western Union nations showed a 'marked lack of appreciation of the value to the common cause of operations in the Middle East'. These forces would be making an essential contribution to the war effort as a whole. Consequently, he continued to believe that nothing should be done to prejudice Britain's agreed strategy of ensuring the safety of the United Kingdom, the Middle East and essential sea communications.[71]

Despite Tedder's continuing commitment to the priorities laid down since 1947 in DO(47)44 by the end of 1949 a discernible shift in CoS thinking was evident. In October the CoS agreed on a major review of Britain's global strategy in which the defence of Western Europe would be given greater emphasis.[72] The first results of this review emerged on 23 March 1950 (Tedder had been replaced by Sir John Slessor) when the CoS finally agreed to put aside their doubts and fears and 'took the decision to commit troops to the continent in the event of war'.[73]

This new approach towards European security was enshrined in a 'Review of Defence Policy and Global Strategy', which was approved by the Defence Committee on 25 May.[74] Sir William Slim summarised the differences between 'The Overall Strategic Plan' (agreed to in June 1947) and the 'Review of Defence Policy and Global Strategy' of 1950 in a note to the Defence Minister, Emanuel Shinwell, on 11 May. He argued that, although there was no fundamental alteration in Britain's overall strategy,

> there was a most important change in the emphasis to be placed on the relative importance of Western Europe and the Middle East. In the past we had been prepared to contemplate the overrunning of Western Europe on the grounds that it would be possible for Britain and the United States to fight back from bases in the United Kingdom and elsewhere. The Chiefs of Staff now considered that the defence of Western Europe must form part of the defence of the United Kingdom. The reason for this change in policy was that it was now considered that, if Europe was overwhelmed, the United Kingdom would be threatened as never before and might well not survive. . . . Loss of Western Europe and loss of the

> Middle East might both mean disaster; but the loss of Western Europe placed us in more immediate peril. The consequence of the policy now proposed would be that the forces available to defend the Middle East at the outbreak of war would be extremely slender. A point in our favour, however, was that the threat to the Middle East would develop more slowly than in the West.[75]

Moreover, it was hoped to get the Commonwealth countries to make a 'substantial contribution to the defence of the Middle East'.

Although the 1947 strategy had been modified in 1948 and 1949, the priority given to the Middle East had remained. With the May 1950 'Review of Defence Policy and Global Strategy' the Middle East remained important but the priority had finally shifted to Western Europe. The debate which had lasted since 1946 was finally decided in favour of a full-blown continental commitment. (This commitment was to be formally emphasised in the Paris Agreement of 1954.)

Throughout the period from 1945 to 1948 British feelings about the defence of Western Europe seemed to have stemmed from a traditional insularity, the kind of mistrust of continentals which Churchill had expressed in 1944. 'The Overall Strategic Plan' also represented a classical British peripheral or maritime strategy in the mode of the Elizabethan or late Victorian strategies. Given Britain's precarious economic position there were strong reasons for avoiding any fresh defence burdens in Western Europe. Allied to these economic anxieties was a growing conviction that as the Cold War gathered momentum Britain's contribution to European defence was of only marginal importance. In any global strategy Britain was better equipped to provide for the defence of the Middle East. During 1948 and 1949, however, it became increasingly apparent that political realities demanded a greater contribution to European defence. Changing perceptions and circumstances brought adaptations in the strategy. Before the British government was prepared to accept such a continental commitment the United States had to be brought more directly into the defence of Western Europe. Without American support both the CoS and the Foreign Office believed Western European security could not be secured. If the 'Commonwealth defence' framework demanded a shift in priority between Europe and the Middle East that could only be achieved through an unprecedented American commitment to European defence.[76]

For Bevin and the Foreign Office, however, the task of securing American military help for the defence of Western Europe was part of a wider 'grand design' of organising the 'middle of the planet'. The Brussels Pact was seen as a key component of the broader Western Union that Bevin had launched in January. To make the Brussels Pact a reality, however, paper guarantees were not enough. Britain had to provide material military support. For the CoS, however, during 1948 broader strategic objectives were more important than Europe and they remained reluctant to plan for an effective continental commitment. As Ivone Kirkpatrick had argued in January, the solution seemed to lie with American support for European defence. How to achieve that American assistance was the key problem after the Brussels Pact had been signed.

7 The Pentagon Talks, 22 March – 1 April 1948

The *coup* in Czechoslovakia on 25 February and Soviet pressure on Norway in early March provided Bevin with the opportunity he was seeking, to encourage greater American involvement in the defence of Western Europe. His approach to the State Department and the Canadian Department of External Affairs on 11 March contained three proposals. The first was a United Kingdom–France–Benelux system with United States backing. The second was a scheme of Atlantic security, with which the United States would be even more closely linked. And the third was a system of Mediterranean security, which would be designed to take care of Italy.[1] Bevin's suggestion that negotiations on his proposals should begin as soon as possible was accepted by Marshall and the Canadian Prime Minister, Mackenzie King. As a result, highly secret talks took place in Washington from 22 March to 1 April during which 'the North Atlantic Treaty was effectively conceived'.[2]

The reason for the secrecy of the talks was partly due to domestic sensitivities in all three countries and partly due to the unwillingness to inform the Soviet Union that such talks were taking place. Despite the growing hostility between the West and the Soviet Union the question of a Western military alliance remained controversial.[3] The Americans also blocked an invitation to France on the grounds that the strong Communist influences in government circles made them a security risk. Gladwyn Jebb, however, has argued that the real reason that the State Department did not want French participation was because 'the French had not been seeing eye to eye of late with the "Anglo-Saxons", more especially as regards Germany, and that it might therefore be better at least to get some sort of understanding with the British on general policy before tackling them'.[4] In the light of later difficulties with the French there was perhaps some justification for this view.[5] Equally some of these difficulties might have been easier to resolve had the French been involved in the discussions at an earlier stage.

Although the initiative for the talks came from the British, each of the delegations arrived at the Pentagon on 22 March with their own proposals on the form an alliance might take.[6]

For the British the objective was to find a way to 'entangle' the United States in the defence of Western Europe. This could either be done by extending the Brussels Treaty to encorporate the United States or by creating a new Atlantic alliance. Britain's main negotiator at the Pentagon talks, Gladwyn Jebb, was instructed to find out first whether the State Department was prepared to accede to the Treaty of Brussels. A better option, however, from the point of view of the Foreign Office, would be to involve the United States in a new scheme of Atlantic security which Bevin had proposed on 11 March.[7]

The Canadian delegation, headed by Lester Pearson (the Under-Secretary of State for External Affairs), had slightly different ideas. Considerable thought had been given to an alternative security system, alongside the United Nations, since the autumn of 1947. On 17 September 1947 the Secretary of State for Foreign Affairs, Louis St Laurent, had suggested in a speech to the General Assembly of the United Nations that the time had come to form 'an association of democratic and peaceloving states willing to accept more specific international obligations in return for a greater measure of national security'.[8] This gave rise to a draft 'Treaty for Greater National Security' prepared by Escott Reid, the Assistant Under-Secretary of State for External Affairs on 4 November. Reid subsequently redrafted this treaty in the light of Bevin's proposals put forward on 11 March. His notion, however, of a world-wide security treaty, including all the British dominions, was not acceptable to Pearson. His view was that the Soviet Union 'would be more impressed by a quick business-like arrangement between UK–US–Canada and France and the Western Union . . . than by an amorphous conglomeration which included Finland, Italy, Portugal and Pakistan'.[9] It was this 'arrangement' which formed the basis of the Canadian proposal at the Pentagon talks.

In contrast to the British and the Canadians, the American position was much less clearly defined. This was largely due to the differences which existed within the State Department on the matter. Some preliminary planning on an alternative security system had been underway since December 1947.[10] The State Department, however, remained divided in March 1948 between John Hickerson's view that the United States should participate in a new Atlantic alliance and George Kennan's opposition to American involvement. The position adopted by the National Security Council and the Joint Chiefs of Staff was somewhere in between. They supported a unilateral guarantee by the United States to the Brussels Pact countries, but no formal treaty of alliance.[11]

Despite these differences concerning the most desirable policy to pursue, 'the American delegation did represent a certain tendency in the State Department'.[12] John Hickerson and Theodore Achilles (the Chief of the Division of Western European Affairs), both supporters of an Atlantic security treaty, were part of the US delegation, while George Kennan was not. He was away in Japan. The position of Hickerson and Achilles was further reinforced by a report of the Policy Planning Staff on 23 March. In this report the PPS proposed the enlargement of the Brussels Pact to take in a number of other European states and 'the creation of a new security treaty to which the United States should adhere'.[13] Consequently, although the differences in the State Department had not been resolved, and were to reappear later, the American delegation at the Pentagon talks did tend to reflect the views of the supporters of an Atlantic pact rather than its opponents.

In the first round of the negotiations, which took place between 22 and 25 March, the delegations discussed four options which might be pursued: a world-wide treaty; the extension of the Brussels Treaty to include the United States and Canada; a unilateral US presidential guarantee to the Brussels Pact countries; and a new Atlantic alliance. By 25 March the first three options had been discarded and there was broad agreement in a joint draft paper drawn up by Achilles, Jebb and Pearson that negotiations should be undertaken on a new scheme of Atlantic security. Despite this consensus, however, differences emerged between the delegations on four main issues relating to an Atlantic alliance. These were the mutual pledge of assistance; the question of indirect aggression: the territorial scope of the treaty; and the prospective membership of the new alliance.[14]

By far the most contentious issue at the Pentagon talks (and later) centred on the question of how binding the mutual pledge of assistance should be. The British delegation, with its determination to secure a firm American commitment to European defence, favoured a strong pledge based on Article IV of the Brussels Treaty. According to this article, all signatories accepted a firm obligation to give each other military and all other aid in the event of an armed attack. Using Article 3 of the Rio Inter-American Treaty as their model, the Canadian delegation suggested a weaker pledge which would oblige members to come to each other's assistance in order to resist an attack. No mention, however, was made of what form such assistance should take. The American delegation adopted the most cautious position of all. In their view each party should agree 'to take armed action' in the event of an armed attack on any of the member states. It should be written into the pledge, however, that every member would determine for itself whether an attack really constituted an armed attack.[15]

In the joint draft paper produced on 24 March the Americans reluctantly accepted assurances from the British and the Canadians, that it went without saying that each party could determine for itself whether an attack constituted an armed attack. It was agreed that this need not be written into the mutual pledge of assistance. Instead the three delegations accepted that members of the alliance would give 'all the military, economic, and other aid' in their power to any of the member states who were subject to attack.[16]

Despite this apparent victory the British delegates recognised that this form of commitment would be very difficult in practice for the American government to accept. The British ambassador in Washington, Lord Inverchapel (who was part of the delegation), sent a telegram to London after the meeting on 25 March, warning that the joint paper 'will on reflexion be considered to be too potent a draft, which even though swallowed, may have to be watered down very considerably'.[17]

Inverchapel's assessment proved to be accurate. After the joint draft paper had been sent to the respective capitals for their comments, negotiations began again on 31 March. This time the Americans insisted on an explicit reference emphasising that every member would be able to determine for itself whether an attack was an armed attack and what kind of assistance to provide. The British and the Canadians felt that they had no alternative but to agree to this change even though, as Jebb argued, it represented 'a considerable watering-down of obligations under the proposed Atlantic Treaty'.[18] The main objective, however, was to secure some form of American assistance and the British delegates kept this very much to the forefront of their minds.

The mutual pledge of assistance, therefore, which was finally accepted reflected the American position rather than that of the British or Canadian delegations. It involved the 'provision that each Party shall regard any action in the area covered by the agreement, which it considers an armed attack, against any other Party, as an armed attack against itself and that each Party accordingly undertakes to assist in meeting the attack in the exercise of the inherent right of individual or collective self-defence recognized by Article 51 of the [UN] Charter'.[19] The key words 'which it considers an armed attack' met the American demand for a clear statement that each state should decide for itself whether an armed attack had occurred. The phrase 'assist in meeting the attack' also removed the obligation to give specifically military support.

A second difference of opinion emerged in the negotiations about whether or not to include a reference in the treaty to 'indirect aggression'. The British delegation was opposed to a provision for consultation in the event of 'indirect aggression', because it might be considered to represent interference in the internal affairs of other states. Inverchapel pointed out to the Americans and the Canadians that the French, who would have to be involved in later discussions, had been 'violently opposed during the earlier stages of the Brussels negotiations to any such thing on the grounds that it would tend to reconstitute a "holy alliance"'.[20]

This view, however, was not acceptable to the other two delegations. The State Department, in particular, was concerned to include a reference to 'indirect aggression' which they defined as 'an internal *coup d'etat* or political change favourable to an aggressor, or the use of force within the territory of a State against its Government by any persons under direction or instigation of another Government or external agency other than the United Nations.'[21] This, after all, had been the device adopted by the Soviet Union in many Eastern European countries. An examination of the paper which was finally accepted by the three delegations reveals that the Americans once again got their way. It was accepted that there should be 'provision for consultation between all the Parties in the event of any Party considering that its territorial integrity or political independence is threatened by armed attack or *indirect aggression* in any part of the world'.[22]

The British delegation also modified its views on two further issues: the territorial scope of the treaty and the question of membership. Ernest Bevin's original idea was to have a treaty which only covered those states bordering the Atlantic Ocean. In the joint draft paper produced on 24 March reference was made to a 'Security Pact for the North Atlantic *Area*'.[23] This implied a broader definition of the territorial scope to be covered by the proposed treaty to include countries which did not border the Atlantic. Disagreement arose, however, over whether to specify the terms of eligibility or not. The Americans were in favour of a precise definition of the geographical area to be covered. The British and the Canadians opposed this on the grounds that a specific delimitation would play into the hands of a potential aggressor. Identifying the area to be included would focus attention, it was felt, on those countries who were not included. Eventually a compromise was found acceptable to all three delegations. It was agreed that the mutual pledge of assistance should cover 'the continental territory in Europe and North America of any Party and the islands in the North Atlantic whether sovereign or belonging to any Party'.[24]

A question still remained over which countries should be invited to accede to the treaty. All the delegations agreed that the United States, the United Kingdom, Canada, France, Belgium, the Netherlands, Luxembourg, Norway, Sweden, Denmark, Iceland and Ireland would be eligible to join. Differences, however, emerged over Switzerland, Portugal, Spain and Italy.

The British were opposed to invitations being sent to each of these, with the exception of Portugal.[25] In their view, despite the ideological problem of Portuguese membership, there were major strategic advantages to be gained from including it in the pact. Switzerland, however, was certain to remain neutral and there was no point in sending her an invitation. Spain had the same ideological difficulties as Portugal, with none of the strategic advantages. And Italy was a Mediterranean country which did not fit naturally into an 'Atlantic' system. Bevin thought it would be better to include Italy either within a Mediterranean system or within the framework of the Brussels Pact once its domestic political difficulties had been resolved.[26]

In contrast the Americans were in favour of Portugal, Spain and Italy while the Canadians were opposed to Portugal and Italy. The Canadians also argued for an accession clause through which 'among others, Western Germany and Western Austria might join'. In the final paper it was agreed to accept this Canadian proposal (but to keep it absolutely secret) and to include Portugal and Italy in the list of possible member states.[27]

The 'Pentagon Paper' which was endorsed by the three delegations on 1 April was therefore a compromise document which in some respects papered over important differences between the United States, Britain and Canada. These were differences which were destined to reappear time and again right up until April 1949. It was also a very sensitive document. The talks had been undertaken in secret without Britain informing the Western European states with whom she had recently entered into an alliance. For this reason the delegates agreed that the 'Pentagon Paper' should be regarded solely as an American document and that the strictest secrecy should apply to it. If, however, the document did leak out it would be possible to argue that it was designed for American planning purposes only and there would be nothing to link it to the security talks from which France and the other Western European states had been excluded. In practice the secrecy was carefully maintained and little was known about the Washington talks from 22 March to 1 April or the 'Pentagon Paper' until the late 1970s. Ironically the Soviet Union probably knew more about the 'Pentagon Talks', through the spy Donald Maclean, than France or the other Brussels Pact states.[28]

The question of whether this secrecy helped or hindered the subsequent negotiations which took place is a matter of interesting debate which will be considered later.[29] Despite the remaining differences, there can be no doubt, however, that the 'Pentagon Talks' played a crucial part in helping to establish the foundations of the North Atlantic Treaty Organisation. In the later negotiations, 'the Pentagon Paper' acted as a kind of secret code which guided the American, British and Canadian representatives. The fact that these three major participants knew more or less exactly what they wanted to achieve was to be of major importance. As Sir Nicholas Henderson (who participated in the later negotiations) has pointed out, the Paper acted as 'an unseen presence, like some new navigational device, to keep the negotiations on a steady course throughout the many months ahead'.[30]

8 The Washington Talks on Security, 6 July 1948 to 9 September 1948

The Foreign Office was highly satisfied with the outcome of the 'Pentagon Talks' but remained very much aware that a great deal had to be done to secure a formal American commitment to Western European defence. On 6 April Bevin still only rated the chances of an eventual agreement with the Americans as little better than fifty-fifty. He told Attlee, 'we shall be lucky if the President and the American Senatorial leaders pronounce in favour of a treaty binding the US for the first time in history to accept positive obligations in the way of the defence of her natural associates and friends'.[1]

Apart from this awareness of the unprecedented nature of such a commitment, Inverchapel was warning from Washington that the pro-pact lobby in the State Department was still encountering stiff resistance, especially from Kennan, who had been absent in Japan during the 'Pentagon Talks'. There was also the problem of Congressional approval. Hickerson had gone out of his way to warn Jebb that Britain must realise that 'some Presidential declaration might in practice be all that the Americans would have to offer. Much would depend on whether some fresh Soviet action maintained the present tense atmosphere. If complete calm prevailed it would be so much more difficult to sell the idea of a pact to Senatorial leaders.'[2] In other words, the pact depended as much as anything else on Soviet behaviour.

Foreign Office apprehension about the American commitment to the Pentagon proposals proved to be justified in the months that followed. Kennan and Bohlen continued to question whether the alliance was necessary. In their view arms deliveries – a military Marshall Plan – would be sufficient.[3] There was also continuing resistance to the idea of a military alliance in the National Security Council and among the US Chiefs of Staff. Instead they favoured negotiations with the Brussels Pact powers on greater defence coordination.[4] This opposition to the notion of an Atlantic pact reinforced doubts in the mind of Robert Lovett, the Assistant Secretary of State, on whether Congressional backing could be achieved. Lovett told Inverchapel on 10 April that he was rather pessimistic about the prospects of the pact because of the 'terrible difficulties in Congress'.[5] At this moment delicate discussions were taking place in the Appropriations Committee on the funding of the European Recovery Program (ERP) which Lovett felt might be obstructed if word got out about possible American involvement in a military alliance. It was also an election year, he said, and there was little the President would be able to achieve in the run-up to November.

Despite Bevin's awareness of the problems in the United States, he was greatly disturbed by Lovett's pessimism and anxious to stop 'the disease spreading.' He therefore decided to change tack. 'Having ridden to a standstill on the Pentagon proposals', he now decided to pursue further initiatives through another channel – the Brussels Treaty apparatus.[6] The other members of the Brussels Pact had been excluded from the secret 'Pentagon Talks' and had no idea what had tentatively been agreed. Bevin now decided to coordinate his efforts with the other Western European states in order to apply more pressure to the US Administration. On 17 April a joint message was sent to Marshall, urging him 'to begin the conversations referred to in previous messages'. There must be no delay, the message said, otherwise 'a favourable opportunity would be missed and a fresh impetus would be given to the cause of communism'.[7]

Marshall was aware of the feeling of frustration which was growing in Western Europe. In his reply to Bevin and Bidault on 22 April he urged them to have no doubt 'as to the intention and determination of this government in relation to the free countries of Europe'. It was important, however, that any assurances from the Administration should have 'maximum country-wide support and backing in Congress'. Nevertheless, he hoped that by the following week he would be able to set 'a definite date' for the opening of negotiations with the Brussels Treaty powers.[8]

Marshall's promise of a definite date raised hopes in the Foreign Office that, at last, further progress towards an Atlantic pact could be made. These hopes, however, were soon to be dashed as weeks went by and no date was forthcoming from the State Department. It soon became clear from 'a routine of regular reconnaissance at the State Department' by British embassy officials that difficulties continued to exist with Congressional leaders and the anti-pact lobby within the State Department.[9] The result was a period of renewed frustration within the Foreign Office over the way the matter was being handled by the Truman Administration. Sir Nicholas Henderson has argued that 'if patience is the proof of diplomacy, the Brussels Treaty Foreign Ministers, and particularly Bevin, were severely tested in the weeks which followed Marshall's telegram on April 22'.[10]

Further progress only became possible when the Vandenberg resolution was passed by the Senate on 11 June by 64 votes to 6. Senator Arthur H. Vandenberg was the Republican Chairman of the Senate Foreign Relations Committee. According to Don Cook, he had all the cliché attributes of a

United States senator. He was a man with 'a large balding head, a bulky torso that overflowed deep leather armchairs, a love of attention, an air of self-confidence, a pompous manner, a constant cloud of cigar smoke, a mellifluous speaking style of loud and florid phrases . . . '. Nevertheless, behind 'this senatorial mien he had a good mind, he was an honest man, he wanted his country to do the right things for the good of the world, and he had a deep love of the Senate, its political workings and the stage that it gave him on which to display power'.[11] He had been sympathetic to the idea of US association with European defence efforts for some time. He was unhappy, however, about the suggestion contained in the 'Pentagon Paper' that, in an election year, the Democratic President should be given such an important part in the negotiations leading to an Atlantic pact. Vandenberg, who had Presidential ambitions of his own, therefore sought a means to turn the Pentagon proposals to the advantage of the Republican party and thereby obtain some personal credit for himself. His resolution was designed to achieve this end.

The resolution urged the US government to pursue certain objectives within the UN Charter, including the 'association of the US, by constitutional processes, with such regional and other collective arrangements as are based on continuous and effective self-help and mutual aid and as affect its national security'. It also called on the US government to contribute to the maintenance of peace 'by making clear its determination to exercise the right of individual or collective self-defence under Article 51 should an armed attack occur affecting its national security'.[12]

The Senate resolution represented an important step forward in providing Congressional support for the idea of an Atlantic pact in some form. It helped to legitimise the debate in the United States during the Presidential election campaign about American support for Europe. The main opposition came from Senator Robert A. Taft and the isolationist wing of the Republican Party, who remained hostile to anything that seemed to be dragging the United States ever more deeply into foreign commitments and European involvements. Despite this limited opposition there was considerable bipartisanship on the issue during the Presidential campaign.[13]

However, problems still existed within the State Department. Kennan and Bohlen continued to have genuine concerns about the long-term effects on East–West relations of permanently dividing Europe by setting up a military alliance. They argued that the pact was provocative to the Soviet Union and unnecessary given the deterrent effect of troops in Europe.[14]

Aware of the influence of Kennan and Bohlen in the State Department debate about American policy towards an Atlantic pact, Bevin felt obliged to address their concerns directly. On 14 May he sent a message to Marshall in which he argued that the case for an Atlantic pact rested largely on its psychological value.[15] Western Europe's main problem was a widespread feeling of insecurity and uncertainty. An Atlantic pact, he argued, would provide the necessary support to bolster Western European governments and help create a greater degree of confidence.

Bevin's argument seems to have struck the right note, at least as far as Kennan and Bohlen were concerned. Although their attitude continued to be sceptical, the following months saw a gradual moderation in their opposition to the idea of an Atlantic pact. Kennan regarded Bevin's letter to Marshall as an 'invaluable contribution' to the debate which was being waged in the State Department.[16]

Despite the impact of Bevin's letter on Kennan, the domestic political situation in the United States still militated against decisive action by the Administration. This was reflected in Marshall's rather discouraging reply to the Foreign Secretary's letter. In the Secretary of State's view there was 'no possibility of completing the necessary negotiations of these matters in time to permit Congressional consideration in the present session'.[17] Bevin took this rather hard and instructed the British ambassador in Washington to explain to Marshall the necessity for holding the talks without delay. The Foreign Secretary was not prepared to sit back until the election was over.[18]

On 14 June Sir Oliver Franks, the new British ambassador in Washington, met Marshall personally to pass on Bevin's message.[19] Franks was a man of brilliant intellect and subtle diplomatic skills who later played a key role in the negotiations which led to the North Atlantic Treaty. At the meeting Marshall said little but indicated that the Administration was finally about to take action. This time the hopes were not to be dashed as they had been on so many occasions in the past three months.

The decisive shift in American policy towards European security coincided with the imposition of the Berlin blockade by the Soviet Union on 24 June. Difficulties had been growing in Berlin since 20 March, when the Soviet military commander in East Germany, Marshal Sokolovsky, walked out of the four-power Allied Control Council set up to administer Berlin and refused to convene further meetings. This was followed on 1 April by a Soviet refusal to allow allied trains through to Berlin. Subsequently, when the trains were allowed to resume, they were subjected to frequent delays and harassment which continued with ever greater intensity until a full blockade was introduced in late June. The decision by the British and American governments to stay in Berlin and to undertake a massive airlift to supply Berlin represented a significant turning-point in the history of the formation of NATO. Don Cook has argued that:

> In those dramatic, history-filled last ten days of June 1948, the hinge of fate turned for the West in postwar Europe. In West Berlin the seemingly irresistible force of Stalinist Communist expansion into Europe met head on with an immovable western determination, whatever the actuality of western power at that time might have been. But this determination in Berlin could never have been mustered without the launching of the Marshall Plan, without growing American involvement in the Greek civil fighting, without the American naval buildup in the Mediterranean, without the signing of the Western Union Treaty in Brussels, without the democratic election victory in Italy, without the diplomatic drive to set a new political and economic course for West Germany, without this gathering sense that the West was moving forward together with a revival of will to restore strength and security. All of this catalyzed in those last days of June around the focal determination: we stay in Berlin. After that the hinge turned.[20]

The United States finally overcame its hesitations over security discussion with the Western European states. On 25 June John Hickerson informed Sir John Balfour, Minister at the embassy, that talks could begin in Washington on 29 or 30 June.[21] After months of frustration patient diplomacy had finally achieved the British government's objective of talks between the United States, Canada and the Brussels Pact powers. Attainment of the eventual goal of American involvement in Western European defence, however, still remained uncertain. A great deal more frustration and irritation was to be felt in the coming months, requiring continuing patience and pragmatism before the Atlantic Pact would become a reality.

After a brief delay, the Washington Exploratory Talks on Security got underway on 6 July.[22] Initially the talks centred on a Six (later Seven) Power Committee of Ambassadors. From the second half of July and most of August, however, more detailed discussions were undertaken in a Working Party consisting of officials from the State Department and the various embassies. As a result of their deliberations the Working Party produced a draft paper in early September which was submitted (via the Committee of Ambassadors) to the seven governments.

Despite the optimism created in Western Europe and Canada by the American decision to call the talks, it soon became clear in the first five meetings that the State Department had no intention of putting forward any proposals of its own. The American position seemed to the other representatives to be like that 'of some modern Minerva, ready to lend its shield to the good cause of European democracy, but not prepared to promise to descend into the earthly European area and become involved itself should trouble occur'.[23] The long shadow of the American election, together with unresolved debates within the State Department, continued to inhibit a more positive American response. Unlike the 'Pentagon Talks' in March, the Washington talks reflected the divisions which had wracked the State Department in recent months. Robert Lovett (the head of the US team), together with Bohlen and Kennan, all continued to have their doubts about the kind of assistance the United States should provide. The Joint Chiefs of Staff and Pentagon officials were also worried that the United States might over-extend itself.[24] At the same time Hickerson and Achilles retained their strong commitment to the idea of an Atlantic Pact. The result was that there was 'no coherent single statement of the US position'. In the early stages of the negotiations Lovett engaged in 'elaborate circumlocutions' which some of the other delegations saw as 'a sinister design aimed at giving the Western European countries the illusion but not the reality of American support'.[25]

Apart from the delaying tactics of the Americans, Sir Oliver Franks, the head of the British delegation, also faced another problem which was to exercise the Foreign Office throughout the negotiations. This involved the approach adopted by the head of the French team, Henri Bonnet. From the beginning, Bonnet argued for the immediate despatch of American military equipment to France and a US guarantee of French territory. The British irritation with what they saw as Bonnet's narrow and selfish approach is summed up by Sir Nicholas Henderson's cutting but amusing portrait of him.

> He was there to state the French case, the whole of the French case, and nothing but it. This he did with remarkable tenacity and tactlessness from the beginning to the end of the negotiations. Not conspicuous for his lucidity of thought or expression, he nevertheless brought with him to the table an array of arresting qualities: an excellent temper, a handsome head, a Maurice Chevalier accent, a gift for generous and irrelevant gesture, and superb pipemanship. As he wielded all these in weary repetition of the French point of view, he appeared quite unconscious of the effect he was having on others. At times his colleagues, particularly the Americans, were to sigh with exasperation, and there were moments when it seemed that Bonnet's importunate behaviour might succeed in breaking everything – but he went on unperturbed.[26]

It was not only the Americans who were exasperated by Bonnet during the negotiations. The Foreign Office files are replete with criticisms of 'the provocative and grasping stupidity of Bonnet' and 'the incredibly stupid behaviour of the French'.[27]

For the British Foreign Secretary the initial meetings of the Committee of Ambassadors therefore raised two important questions. How to persuade the French to change what was regarded as their shortsighted attitude in the interests of Western European security as a whole; and, secondly, how to get a more forthcoming attitude from the American delegation (despite the caution arising from the approaching election) so that some progress could be made before the end of the year. These twin aims became the core of British diplomacy in the months that followed.[28]

The problem of securing a more responsive approach from the United States during the negotiations was complicated by the difficulties which had arisen in Anglo-American relations in the years since the end of the war, especially over Palestine. Escott Reid has shown that the task of reaching an agreement during the negotiations 'was more difficult than it otherwise would have been . . . because during 1948 differences over Palestine strained relations between the governments of the United States and Britain'.[29] In November 1947 the United Nations voted in favour of a US-sponsored resolution on the partition of Palestine. Britain abstained on the issue. It was not long before bitter fighting broke out between Arabs and Jews in Palestine which continued for the rest of the year. This brought recriminations at the highest level of government in both Britain and the United States. Bevin thought that Truman had supported Zionism because of his desperate need for 'Jewish votes, money and influence in the Presidential elections in November'. Truman, on the other hand, felt that the British Foreign Secretary was deeply prejudiced in favour of the Arabs against the Jews.[30]

Not only did these differences colour the general conduct of the negotiations over the North Atlantic Treaty but it seems that a direct attempt was made by the Americans to link the two issues together directly. Dean Acheson, who was Director of the Office of United Nations Affairs in the State Department, sent a message to the Canadian government arguing that 'an understanding over Palestine was necessary if the security pact was to be concluded'. It can be assumed, as Reid has argued, that 'the United States administration gave this warning directly to the British government'.[31]

Despite this complication, the central and single-minded aim of the British delegation at the Working Party meetings which began on 14 July was to secure the involvement of the United States in a regional arrangement for collective self-defence. It did not really matter what form the arrangement took: 'the main thing was to secure US participation'.[32] From the British point of view it was important to make sure that the problems over Palestine did not prevent the achievement of this objective.

As the discussions in the Working Party progressed, it appeared to the British delegation that a more positive American attitude was emerging. On 10 August Lovett 'gave a flying start to the discussions by circulating a paper based on the Rio Treaty'.[33] There was still no formal commitment from the United States but it seemed to the British representative on the Working Party, Derick Hoyer-Millar, that Kennan and Bohlen were finally coming round to the idea of a treaty.[34]

Just as British hopes were rising, however, an intervention by Henri Bonnet set back the steady progress which was being made. Towards the end of August, Bonnet suddenly informed Marshall that the French government could only pursue the idea of a North Atlantic Pact *provided* they got satisfactory assurances on four points: (1) that there would be an immediate dispatch of American military supplies to France; (2) that US troops would be immediately sent to France; (3) that immediate arrangements would be made to set up an integrated military command in Western Europe; (4) that French representatives would be allowed to join the Combined (Anglo-American) Chiefs of Staff organisation.[35]

Presented as they were on 'a take it or leave it' basis, Bonnet's representations 'had the worst possible effect on Marshall'. The Secretary of State was an open-minded, undogmatic man and the French ultimatum was wholly unacceptable to him. He was 'outraged' by the presumption that France should lay down conditions for joining the Pact. The French proposals also had an adverse effect on Kennan and Bohlen. They began to question once again whether a pact was such a good idea. Despite his recent conversion, Kennan took the view that 'if the French do not want a treaty, we had better drop the idea after all'.[36]

Derick Hoyer-Millar, Minister in the British Embassy in Washington, expressed his fear to the Foreign Office about the effect Bonnet's approach was having in Washington. There was a danger, he argued, that 'he might jeopardise not only the French interests but those of other powers which were struggling for a treaty'.[37]

The vehemence of this and earlier criticisms of the French were perhaps understandable but not altogether fair. France was, after all, excluded from the Pentagon Talks in March 1948 and as a result had no knowledge of the 'Pentagon Paper' which formed the basis of the British, American and Canadian negotiating positions during the Washington Talks.[38] The French believed that they had been invited to increase European security. For France this meant the opportunity to secure American aid. Instead, the discussions centred on the idea of a pact which came 'as a complete surprise to the French government'.[39] As a result they became suspicious and preferred to put their case forcefully for concrete aid rather than accept 'proposals of a vague and general character'.[40]

It appears that the long and acrimonious disputes which followed were largely the result of what Gladwyn Jebb called 'a genuine misunderstanding in Paris about the objectives of the Washington Talks.'[41] Bonnet's rather intransigent diplomatic style did not help the progress of the negotiations, but this can, in part at least, be explained by the French feeling of disappointment when they discovered that the aid they thought had been promised was not going to be forthcoming.

For the British, however, in the summer of 1948, the singleminded approach adopted by the French jeopardised one of the central objectives of British foreign policy. As a result of Hoyer-Millar's warning, therefore, the Foreign Office embarked on a concerted attempt to modify the French position in order to make it 'more realistic and constructive'. On 23 August Jebb went to Paris to visit Jean Chauvel, the Secretary-General of the French Foreign Office, and Robert Schuman, the new Foreign Minister. Jebb regarded the visit as being 'most fruitful'.[42] He succeeded in persuading the French of the need to distinguish between immediate and long-term security problems. The French government accepted the main lines of the US approach to a North Atlantic Pact and agreed to send new instructions to the French embassy in Washington to pursue a more constructive approach.

As a result, despite some continuing difficulties, the Working Group managed to produce a paper for the Committee of Ambassadors at the beginning of September which represented 'an agreed statement on the nature of the problems discussed and the steps that might meet these problems'. The paper, which closely resembled the earlier secret Pentagon proposals, was subsequently sent on to the seven governments for their consideration.[43]

For the British, despite a number of continuing differences of opinion on the nature of the mutual guarantee and membership of the alliance, the central objective of British diplomacy over the past nine months appeared to have been virtually achieved. Jebb wrote to Bevin on 9 September to inform him that 'at this stage . . . . it can be said that the probability that America will in fact now enter some system for the defence of western Europe is very considerable, and that failing some new and rather unexpected development it is likely that this policy would be continued by a Republican administration'.[44] Although Jebb's optimism was well-founded, a number of frustrating events were still to arise before the North Atlantic Pact became a reality.

9 Last-Minute Problems, 9 September 1948 – 28 March 1949

After the American election was over and Truman was returned to office, the Committee of Ambassadors resumed negotiations on the reactions of the seven governments to the September Paper. The most notable feature of the talks which began on 10 December was a much more positive American attitude than in the past.[1] Lovett, in particular, was now determined to make rapid progress towards an early signature of the pact. This contrasted with the earlier American approach. Now, instead of 'dragging their feet', the US delegation were anxious to make swift progress and finalise a draft treaty before the end of the year.

This change in American attitudes was reflected in the support George Kennan was now prepared to give to the idea of an Atlantic Pact. Kennan still had some of his earlier reservations but he now believed that 'there was a valid long-term justification for a formalisation, by international agreement, of the national defence relationship among the countries of the North Atlantic community'.[2] In a staff paper written on 23 November he argued that 'the conclusion of such a pact is not the main answer to the present Soviet effort to dominate the European continent and will not appreciably modify the nature or danger of Soviet policies'. He conceded that a military danger continued to exist and appeared to be increasing but, in his view, 'the basic Russian strategy was the conquest of Western Europe by political means'. He remained convinced that 'if war came in the foreseeable future, it would probably be one which Moscow did not desire but did not know how to avoid. The political war on the other hand, is now in progress; and if there should not be a shooting war, it is this political war which will be decisive.'[3]

A NATO Pact, Kennan argued, would affect the political war only in so far as it increased the self-confidence of Western Europe in face of Soviet pressure. He still believed that the need for military alliances and rearmament on the part of the Western Europeans was 'primarily a subjective one, arising out of their own minds as a result of their failure to understand correctly their own position'. Their best course of action, if they were to save themselves from Communist pressures, remained the struggle for economic recovery.

Nevertheless Kennan accepted Bevin's earlier argument that, whether or not the Pact was necessary objectively, if it did help (as he thought it would) to boost Western European confidence then it was needed and desirable. He therefore urged that a NATO Pact should be concluded as soon as possible.

With the election over and Congress as well as public opinion in favour of the Pact, the earlier anxieties in the State Department were swiftly evaporating. As a result, even the discussions on the mutual guarantee clause, which had proved so difficult in the past, were, initially at least, much smoother than expected.

Following the problems which had arisen over the mutual assistance pledge in the 'Pentagon Paper' in April, the United States delegation in the Washington Talks remained concerned about two specific issues.[4] The first was the question of whether the pledge should state explicitly that each ally had the right to decide for itself whether an armed attack had occurred. The second was whether the pledge should make an explicit reference to the provision of military assistance if an ally was attacked. On both issues Britain and the other states argued in favour of a strong pledge. They remained opposed to explicit statements about states having the right to decide whether an armed attack had occurred. This they believed remained self-evident. At the same time they contended that it was essential that reference must be made to military assistance if the treaty was to achieve its purpose of deterring aggression.

By 24 December the United States government had accepted these arguments.[5] In the draft treaty which was drawn up at this time to be submitted to all the governments concerned it was agreed that the pledge should read:

> The Parties agreed that an armed attack against one or more of them occurring within the area defined below shall be considered an attack against them all; and consequently that, if such an armed attack occurs, each of them, in exercise of the right of individual or collective self-defence recognized by Article 51 of the Charter of the United Nations, will assist the party or parties so attacked by taking forthwith such military or other action, individually and in concert with the other Parties, as may be necessary to restore and assure the security of the North Atlantic area.[6]

The British government viewed this wording with great satisfaction. In essence it was much the same as the pledge contained in the 'Pentagon Paper' in April and was 'as strong or stronger than that in the Brussels Treaty'. With the United States apparently committed to a firm guarantee of support and most of the other articles agreed in the draft-treaty of 24

December the rest appeared plain sailing. This was not, however, to be the case.[7]

January was to bring new last-minute difficulties. With the retirement of Marshall (and Lovett), Dean Acheson became Secretary of State. Initially Acheson was preoccupied by a number of urgent issues (particularly over Palestine), which required his immediate attention. When he eventually turned to the North Atlantic Pact in early February he found that only general agreement had been reached with Congressional leaders. Acheson therefore felt obliged to consult Senate leaders, before making further progress. As a result the Brussels Pact powers were once again forced to remain patient while discussions which they assumed had been completed were resumed between the Administration and Congress.[8]

British officials were surprised to find that the wording of Article 5 on the mutual guarantee, which represented the cornerstone of the Treaty, was still a bone of contention in some quarters in the United States. The belief was that the issues had been settled in the Paper of 24 December and the subsequent discussions with Lovett and Bohlen in January. Reports, however, were received in early February that Senator Connally, the new Democratic Chairman of the Senate Foreign Affairs Committee had raised new objections to the mutual guarantee clause.[9] On 14 February Senator Connally spoke in an impromptu debate on the Atlantic Pact and criticised what he saw as the automatic commitment to go to war contained in Article 5.

Connally's speech was badly received in the Foreign Office. The Senate debate was described as 'deplorable'.[10] Connally was accused of acting from 'pique' because he had not been consulted as much as Senator Vandenberg by the Administration.[11] The situation was made worse by Vandenberg who once again argued that it must be made clear in the pledge that 'it was up to each party to determine the nature and character of the assistance' it would give to an ally.[12]

Faced with this opposition Acheson proposed that Article 5 should be changed. At a meeting of the Ambassador's Committee on 8 February he suggested that the phrases 'military and other action' and 'as may be necessary' should be deleted.[13] In their place the pledge should simply read 'to take action and assure the security of the North Atlantic area'.[14] He was supported in this even by Hickerson and Achilles who felt that the Administration would have to give way to the demands of the Congressional leaders.

Hickerson and Achilles, however, persuaded Acheson to discuss the matter with Oliver Franks, the British Ambassador, to get his views on whether the changes would impair the objectives of the Treaty. Franks told the Secretary of State that the 'watering down' of the pledge would be 'disastrous'.[15] In his view there had to be some reference to military assistance. This was also the view of the British Foreign Secretary. On 17 February he sent a long, trenchant and unusually irritable telegram to Acheson in which he said that he was

> seriously disturbed at these developments. Having regard to all the conversations and negotiations over this pact which we have held with the US government during the last year, we have been basing our policy on the confident expectation that the US was willing to join in the creation of a solid Atlantic Community capable of resisting aggression from any quarter. I thought the principle had been accepted and there was no difference of status between the various parties in such a unit (*except indeed that we, the European members, were in the front line and would probably take the first knock in any emergency*). We were willing to take the risks which such a pact involved for us on the assumption that the obligations were equal all round. This was the position we ourselves accepted when we asked our continental friends to accept the risks involved in the Brussels Treaty.[16]

Bevin went on to argue that the grave doubts which had emerged over the US attitude could well cause 'every European country to shrink from the risks which they would be exposed to by signing any pact'. At the end of the telegram his frustration with yet another delay led to a veiled threat that Britain might have to reconsider its position on the pact. Britain, he said, would want to know 'whether or not the US government accepted the conception of the Atlantic Community as one fundamental unit which has got to be defended together for the sake of all the parts', before the government would be willing to sign the pact.[17]

In the event Bevin's worries proved to be shortlived. The problem was finally settled when Acheson, with firm backing from Truman, persuaded Connally to agree to a compromise which was acceptable to him and to the Brussels Pact Powers on the wording of the mutual guarantee clause.[18] Largely because of the election campaign, Truman's role in the negotiations had been limited. Like Attlee, he provided support for the officials but allowed them to deal with the detail of the negotiations without undue interference. On this occasion, however, he involved himself directly to overcome the dangerous impasse which was developing. When Acheson went to see him at the Oval Office on 17 February to discuss the problem with Connally, the President told his Secretary of State that it was absolutely essential to include a commitment to military action in the treaty. He also promised to ring Connally personally and invite him to Blair House to discuss the matter. Whether this meeting ever took place is not clear but according to Don Cook, 'The result of this prudent intervention by the President is clear. Harry Truman's leadership preserved a viable North Atlantic Treaty.'[19]

The inclusion of the phrase 'action, including the use of armed force' satisfied Britain and the European states that they had achieved as much as they were likely to get from the Americans on a commitment to use force. At the same time the inclusion of another phrase 'as it deems necessary' met the senators' concern about the need to emphasise the right of each nation to decide what action to take in the event of aggression.

If the British were reasonably successful in getting their way over the mutual pledge clause they completely failed over another dispute which came to a head in the first few months of 1949. This centred on the provision in the treaty for economic, social and cultural cooperation.[20] The Canadians were determined that the alliance must be more than just military in character. It should emphasise the spiritual crusade of Western values against Communism. On their initiative the 'Pentagon Paper' of 1 April had contained an article urging members of the proposed alliance to make 'every effort, individually and collectively to promote the economic well-being of their peoples and to achieve social justice in order to create overwhelming moral and material superiority as well as military superiority, in the cause of peace and progress'.[21]

They were supported in this proposal by the United States and, in the Washington Talks, Article 2 was framed to emphasise the importance of economic and social cooperation. The British, however, were strongly opposed to this Article and argued that it should be deleted. Jebb told the Working Group on 9 September 1948 that Bevin

> had expressed considerable concern over the emphasis being placed in these talks on the establishment of machinery for the solution of common economic and cultural problems. In his view, this would only duplicate much of the machinery now in existence, such as the OEEC, but might inject considerable confusion into the international picture and slow up the present progress of the European nations toward that union which they all believe is so essential.[22]

For a time in November Britain managed to persuade the other Brussels Pact states to go along with their opposition to Article 2. This support for Britain soon dissolved and it was accepted as part of the draft treaty agreed on 24 December. However, the issue reemerged in February when the new Secretary of State reversed the American position and began to oppose the inclusion of Article 2. Acheson argued that there was little support from leading senators for broadening the provisions of the treaty.[23]

This important change in the American position was greatly resented by the Canadians. At the end of February they went as far as to tell the Americans 'that unless we get an article on these lines in the treaty the Canadian government would have to review its position towards the whole project'.[24] This was a polite way of saying that they regarded Article 2 as so important for domestic political reasons that they would refuse to sign an Atlantic treaty unless it was included.

At this point, after an appeal by the Canadians to the British government, Bevin modified his longstanding objection to the article. Reluctantly he agreed to accept Article 2 not because Britain 'had any special interest . . . in having such a provision in the treaty' but rather to meet Canadian wishes on the issue.[25] The Canadians obviously felt that the issue was so important that the British government was reluctant to jeopardise the treaty at this late stage by continuing their opposition to what was regarded as a minor issue.

By late February Canada had received the support of all the Brussels Pact powers and Acheson finally accepted the consensus in favour of including Article 2. According to this article the Parties agreed:

> to contribute toward the further development of peaceful and friendly international relations by strengthening their free institutions, by bringing about a better understanding of the principles upon which these institutions are founded, and by promoting conditions of stability and well-being. They will seek to eliminate conflict in their international economic policies and will encourage economic collaboration between any or all of them.[26]

No sooner had the problems of the mutual guarantee clause and Article 2 been settled than a new difficulty arose, this time with the French. Once again the Foreign Office believed that French behaviour was threatening the success of the negotiations.[27] Throughout the negotiations differences had existed over which countries should be invited to join the Pact and the area covered by it. The French had insisted all along that Algeria should be included in the treaty area. Most of the other states had resisted this on the grounds that it set a precedent for other colonial areas and posed the risk of drawing them into colonial conflicts.[28] On top of this dispute, the French suddenly demanded Italian membership of the Pact and indicated that if their proposals were not accepted they would have to reconsider their whole position on the alliance.[29]

Britain was very much opposed to Italian membership on the grounds that it was not an Atlantic country and could be a liability to the alliance because of its domestic difficulties.[30] It was also believed that 'Italy would demand as its price for entering the alliance far-reaching changes in the Italian peace treaty, in particular the return of some of its colonies'. The Foreign Office, however, was worried once again by the intransigence of the French and the impact which this was having on the Americans.[31]

By 1 March Britain was prepared to modify its position on Italian membership. Franks told the other Ambassadors that 'the UK had always thought that on the whole it was better not to have Italy as a member of the Pact. However, . . . the strength of that preference had diminished during the progress of negotiations. The UK would not stand in the way of a general opinion and would not press its initial preference.'[32] If there was a consensus in favour of Italian membership Britain would accept it.

Despite the irritation caused by renewed French threats not to sign the Pact, Britain was also prepared to accept the inclusion of Algeria within the treaty area. Bonnet had argued that Algeria was an integral part of France. From the French point of view it was 'in the same relation to France as Alaska or Florida to the United States'.[33] Algeria therefore had to be included in the territory covered by the pledge. Britain went along with French demands even though the United States and Canada opposed the inclusion of Algeria on the grounds that the alliance might be dragged into colonial disputes. Both countries were also worried by the domestic reaction to the inclusion of colonial territories in the area covered by the treaty.[34]

The final few weeks of the negotiations brought a bitter dispute between France and the United States over the Italian and Algerian issues. By late January it might have been possible to diffuse the problems associated with Italian membership and whether the pledge should cover Algeria. The French had apparently championed Italian membership as a bargaining tactic to secure greater support for Algeria.[35] By 24 January the United States had moved to a position in its own deliberations in which it was prepared 'to swallow Algeria'.[36] If the American government had announced this at the time it might have been possible to have arrived at a compromise agreement with France which included Algeria but excluded Italy. In practice, however, the United States let the issue of Algeria 'hang in the balance' for a further five weeks until they finally accepted the inclusion of the three Departments on 1 March.[37] The result was a prolongation of the dispute and in the end both Italy and Algeria were included.

In the final weeks of the negotiations American diplomacy appears to have suffered from a certain lack of consistency. On the question of Italian membership the United States changed its position on a number of occasions. Sometimes it was in favour of Italy joining, sometimes it was opposed. This was largely the result of differences within the State

Department between Hickerson and Kennan. John Hickerson consistently championed Italian membership on the grounds that 'an Atlantic Community that did not include Italy would not only be incomplete but contrary to our heritage'. He believed that Italy would almost certainly go Communist if it was left out of 'a group to which historically and culturally it belongs'.[38] He was opposed in this by George Kennan who believed that, if Italy was included, Greece, Turkey and other countries might claim that they also had a case to join. The commitment would then be extended and diffused. As a result of the difference between the two officials it was very difficult for the other delegations in the Washington talks to know exactly where the United States stood on the question of Italian membership.

The confusing nature of the American attitude towards Italian membership is summed up by Escott Reid. Hickerson, he argues, was the grey eminence in the story.

> From the beginning, he wanted Italy in the alliance; and whenever he lost the argument within the State Department, he bided his time and returned to the attack. He was forced to agree to a compromise in September of 1948; he won in November, but then had once again to agree to a compromise in December. In the *de facto* interregnum between the Lovett regime in the State Department and the Acheson regime, he committed the United States to strong support of Italian membership. Acheson the next month withdrew this support. Then, presumably under pressure from Hickerson, Acheson managed to convince the President, the senators and the governments of Britain and the Benelux countries to acquiesce in Italian membership.[39]

Apart from internal differences within the State Department the consistency of American diplomacy in the final stages of negotiations was also affected by a deep resentment over French tactics. The attempt by France to link the Italian and Algerian issues was regarded by the State Department as nothing short of blackmail. Their irritation was shown when they decided to adopt a unilateralist approach to prevent further delay. Without consultation with any of the other states involved, Acheson decided to invite Denmark, Portugal, Iceland and Italy to join Norway in the discussions with the other seven powers before the treaty was signed.

From the British viewpoint this American approach was almost as high-handed as that of the French. The Americans had 'jumped the gun' and faced the other parties with a *fait accompli*. The general view in the Foreign Office was that the State Department 'have certainly treated us all with scant courtesy'.[40] Unlike the French, however, whose behaviour delayed the process of negotiations and threatened the pact itself, the Americans were at least seeking an early signature of the treaty. The Foreign Secretary therefore was 'not disposed to take very strong exception'. There was nothing Britain or the others could do about it. Consequently, it was better 'to swallow the pills smilingly' in the interests of the wider objective of entangling the United States in European defence.[41]

The final stages of the negotiations in March tidied up various loose ends. Despite the misgivings which some of the original seven participants felt about inviting Denmark, Portugal, Iceland and Italy to join the negotiations (together with Norway), no significant new proposals were raised at the last minute. Anxiety over Italian membership receded when the Italian government gave assurances that no attempt would be made to involve its new allies in the Trieste question or any of the residual issues associated with the former Italian colonies. Portugal was happy to accept the invitation to join after obtaining the blessing of the Franco regime in Spain. Denmark decided to follow Norway into the alliance rather than opt for neutrality. (With Danish membership the United States and the alliance achieved two important strategic aims of securing base facilities in Greenland and ensuring control of the Baltic States.) Despite left-wing criticisms, Iceland also felt that its security was more likely to be assured inside the alliance than outside.

The only remaining issue to resolve concerned the boundaries of the treaty area. After the decision to include 'the Algerian Departments of France' in the treaty area the main outstanding problem centred on the designation of the northern and southern boundaries. It was finally agreed that in the north the treaty should cover the whole area to the North Pole – 'well beyond the Norwegian-controlled island of Spitzbergen and beyond the Northwest Territories of Canada'. As regards the south, an American suggestion was accepted that the Tropic of Cancer should act as the boundary. This allowed the United States to make a clear distinction between its two main postwar treaties – the NATO Treaty and the Rio Treaty.[42]

On 15 March the Ambassadors' Committee met for the last time to give their final approval to the draft treaty. In so doing, despite the difficulties of the past three months, they brought to a successful conclusion eight months of negotiations which began as the Soviet Union imposed its blockade of Berlin. At the beginning the Western European states (and the United States) had been uncertain how far the United States would go in reversing its traditional reluctance to become involved in 'entangling alliances.' By the end of March 1949 the United States had demonstrated through its decision to stay in Berlin and the leading role it played in the negotiations that a new era in American foreign policy had dawned.

10 Conclusions and Achievements

On 4 April 1949 the North Atlantic Treaty was signed in Washington. For Britain the signing represented the successful completion of a major goal of foreign policy, the origins of which can be traced back to wartime planning. It would be wrong, however, to conclude that Britain pursued a wholly deliberate and consistent policy from 1944 onwards which culminated in the establishment of the North Atlantic Treaty Organisation. The wartime and postwar periods were times of great uncertainty, and different visions of the future had to be constantly modified in the light of changing circumstances.

After Trygve Lie's initiative at the end of 1940 a great deal of planning for the postwar period took place in the Economic and Reconstruction Department of the Foreign Office and the inter-departmental Post-Hostilities Planning Staff. During this process two rather different ideas emerged about European security following the end of the war. The idea favoured in the Foreign Office was that Britain should play a leading role in coordinating the policies of the Western European states. This was reflected in Gladwyn Jebb's emphasis on a Western European group. This would allow Britain to remain a Great Power in the postwar world and achieve a leading role in Western European affairs. The Chiefs of Staff, on the other hand, with their greater suspicions of the Soviet Union, argued that a Western European group would not be sufficient to provide security if great-power cooperation broke down – which they believed it would. What was needed was a continuing close military relationship with the United States which had been so crucial during the war. These differences of emphasis were reflected in the disputes between the Foreign Office and the Chiefs of Staff over the two PHPS reports on 'Security in Western Europe and the North Atlantic' produced in July and November 1944 and the global survey undertaken from late 1944 through to the summer of 1945.[1]

The failure to resolve the dispute continued into the postwar period. In August 1945 Ernest Bevin declared that his long-term objective was extensive political, economic and military cooperation throughout Western Europe, starting with an Anglo-French alliance. This 'grand design' fitted in with Foreign Office wartime and postwar planning, which was concerned to establish a Western European group to achieve an independent role for Britain. When the Chiefs of Staff warned that an Anglo-French alliance should only be concluded provided it did not damage Anglo-American relations, the Foreign Office were determined to go ahead irrespective of the risk. Sir Orme Sargent, the Permanent Under-Secretary at the Foreign Office, argued in December 1946 that if Britain made every move in the realm of high policy contingent on American prior approval, the 'prospects of being able to give a lead to Western Europe will vanish and we shall never attain what must be our *primary objective* viz by close association with our neighbours to create a European group which will enable us to deal on a footing of equality with our two gigantic colleagues, the USA and the USSR'.[2]

With Britain's desperate economic problems in 1946 and 1947 the Foreign Office saw a Western European grouping (supported by the Dominions and the United States) as the main vehicle in the long term for helping Britain overcome its economic problems and reestablishing its great-power status. Economic considerations were perhaps the most important part of Bevin's foreign policy towards Western Europe in this early postwar period.[3] The pursuit of 'the primary objective' of a Western European group, however, was complicated by the rapidly-changing circumstances. The scale of the economic problems facing Britain required immediate assistance, which could only come from the United States. American support was also necessary as relations with the Soviet Union deteriorated. This meant that 'too great independence from the United States would be a dangerous luxury'.[4] At the same time the task of securing an Anglo-French alliance as the cornerstone of a Western European group proved difficult to achieve due to differences over the Levant and the Ruhr.

When the opportunity arose to conclude a treaty with France, expediency was as important as the pursuit of a 'grand design'. Bolstering democratic socialism in France coincided with the broader objective of Western European unity. The same pragmatism was evident when Bevin postponed an approach to Belgium and Holland to conclude treaties of alliance, in order to respond to Marshall's offer of aid on 5 June 1947. The Foreign Secretary also demonstrated considerable diplomatic flexibility in changing his emphasis in early 1948 from bilateral treaties, based on the Dunkirk Treaty, to a broader Western European pact when Belgium and the United States advocated a multilateral alliance framework. As time went on, however, Bevin began to see that Western Europe could not be the predominant focus for achieving Britain's economic, political and strategic interests. In a paper prepared for the Cabinet in January 1948 the Foreign Secretary expanded his conception of the Western Union. He argued that Britain 'should seek to form with the backing of the Americans and the Dominions a Western democratic system'.[5] Bevin's conception of a Western Union developed in this paper referred to Europe not only as a geographical conception, but also argued that Europe had extended its influence throughout the world and consequently it was necessary 'to look further afield' in the first place to Africa and South East Asia. There was a need, he argued, to mobilise the resources of Africa and other British and European colonial territories. 'If those were included the whole would form a bloc, which in population and productive capacity, could stand on an equality with the Western hemisphere and the Soviet blocs.' The pursuit of British leadership in Western Europe remained an important objective but that had to be only part of a broader international framework for the pursuit of British interests.

In some important respects this brought British foreign and defence policy closer together in early 1948. Bevin's expanded conception of the Western Union fitted in with the Chiefs of Staff planning for a 'Commonwealth defence policy'. Even though Bevin was thinking of a coordination of democratic states particularly in the political and economic spheres, he was not averse to the coordination of defence policies as well, especially as relations with the Soviet Union deteriorated. Difficulties continued, however, between the military chiefs themselves and between the CoS and the Foreign Office over the relative priority which should be given to different components of Britain's global security framework.

Despite economic difficulties, the main thrust of British defence policy in the immediate postwar period was to maintain the necessary military capabilities to allow Britain to continue to play a global role. This involved giving priority to nuclear weapons and the search for a global security system based on the Middle East and embracing the Dominions and the United States. Continuing attempts were made to improve cooperation with the Commonwealth countries which began to bear fruit from 1948 onwards. The central objective of military planners, however, was to ensure the continuation of close military ties with the United States in all areas where British security interests were threatened by the Soviet Union. It was in this context of a clearly-perceived Soviet threat, the requirement for global military capabilities and the crucial importance of a close defence relationship with the United States, that the CoS viewed British policy towards Western Europe. This approach to global strategy was clearly spelled out in the 'Overall Strategic Plan' in June 1947.

With the severe economic pressures on the defence budget, all the Defence Chiefs initially agreed that despite the political commitments being made to Western European states, the military capabilities were not available to sustain a long military campaign on the continent. This was not, however, a position which was held with equal conviction by all the Chiefs of Staff. In 1946 Montgomery was arguing that more attention would have to be given to some form of continental commitment. The issue, however, only came to a head in early 1948 when the Chief of the Imperial General Staff urged the Naval and Air Chiefs to provide greater military support to back up Britain's growing political commitment to the Western European states. Montgomery was supported in this by the Foreign Secretary, Ernest Bevin. Tedder and Cunningham, on the other hand, continued to argue that Western Europe was indefensible without American assistance and that Britain should avoid making a major contribution to continental defence. In this they received support from the Prime Minister, Clement Attlee. It was not until 12 May 1948 that Montgomery secured the reluctant agreement from the Defence Committee that Britain should stay and fight on the Rhine if war broke out. Even then the CoS continued to refuse to plan to send reinforcements to the continent in the event of conflict. The continental commitment remained limited until March 1950 when significant adjustments were made in British strategic planning to put more emphasis on the defence of Western Europe.

We see here something of a contradiction in British foreign and defence policy in the late 1940s. The Foreign Secretary remained intent on playing an important role in Europe as a cornerstone of his Western Union policy while the Chiefs of Staff were unwilling to provide the military forces to fight on the continent in the event of war. The Middle East remained a more important focus in British military planning. The problem which this created for the coordination of British diplomatic and military policies was stressed by Ernest Bevin in a meeting of the Defence Committee on 8 January 1948. He emphasised the 'embarrassment which would be likely to arise with our potential European allies if our plans were based on the assumption that we did not intend' to provide military forces to defend the continent.[6] This point was reiterated the following day by Ivone Kirkpatrick when he argued that, if the policy of limited liability towards continental defence was maintained, Britain would eventually either have to admit to its allies that it had no plans to come to their assistance despite treaty guarantees or refuse to disclose its intentions.[7] As time went on, the allies pressed for precise information about Britain's strategic plans and the reluctance to disclose information about military policy undermined Britain's leadership aspirations in Western Europe.

During 1948 and 1949 the contradiction between the different strands of policy became increasingly evident. Britain saw American assistance as a solution to the problems caused by its reluctance to provide military support to its European allies. The United States, however, itself remained reluctant to support European defence at least until the Brussels Pact powers had demonstrated a concrete willingness to defend themselves. Even then the form of American military assistance to Western Europe continued to be unclear.

With the *coup* in Czechoslovakia, Soviet pressure on Norway and the onset of the Berlin Blockade, the Foreign Secretary believed there was 'a threat to Western civilization'. Security considerations became paramount for the British government and the Foreign Secretary's major preoccupation shifted from the task of building a Western Union to reinforce Britain's independent power position to a more permanent Atlantic Alliance. Elizabeth Barker has argued that 'the signing of the Atlantic pact fulfilled Bevin's dream of harnessing America's military power and resources to the defence of Western Europe . . . But at the same time it marked the . . . end of Bevin's other dream of British leadership of the Western Europeans.'[8] Likewise it signalled the end of British ambitions for a more independent world role.

The Atlantic strand of policy finally emerged victorious over the European and 'third power' strands. In early 1949 this became evident when the British government decided to give priority to securing 'a special relationship with the United States and Canada . . . for in the last resort we cannot rely upon the European countries'.[9] The CoS had expressed this view consistently since 1944 and increasingly Bevin had come to accept it during 1948. Although the need for close Anglo-American relations had been accepted by the Foreign Office throughout the postwar period this had been seen initially largely as a temporary measure to help reestablish a more independent role for Britain. That view gradually disappeared during 1948, and by 1949 it was recognised that a 'special relationship' with the United States was a desirable longer-term goal. The 'third power' role was now seen as being unrealistic.[10] In February Bevin set up the Permanent Under-Secretary's Committee 'to consider long-term questions of foreign policy and to make recommendations'. In one of their first reports in March, entitled 'Third World Power or Western Consolidation', they emphasised that sound Anglo-American relations should be the cornerstone of British foreign policy. According to the report, 'in the face of implacable Soviet hostility and in view of our economic dependence on the United States . . . the immediate problem is to define the nature of our relationship with the United States'. It was argued that Britain faced a choice between trying to establish a third world power based on Europe and the Commonwealth, or developing a union with the Atlantic area states 'in a Western preponderance of power' which could deter the Soviet Union. In practice, the paper argued, the option of setting up a strong independent third world power was not available. The Commonwealth was not 'a geographical or strategic unity' and an effective united Western Europe was not a practical possibility because of strong national traditions and differences.[11]

Germany remained another problem for a 'third power' based on Europe. 'To be really strong enough to maintain itself', it was argued, a Western European 'third power' must include 'a strong and probably united Germany'; but in that case Germany must either attempt to dominate the union or to use its position to levy blackmail from Western Europe as the price of abstaining from cooperation with Russia.[12] Added to this was the fact that the eventual economic viability of Western Europe was still uncertain. The initiative for economic cooperation had come from the United States. If this assistance was withdrawn, European cooperation might stall. Even if it continued, it was 'far from clear that integration would confer any greater power of united resistance'. The paper also went on to argue that military cooperation between Western European states was unlikely to provide an effective defence. Unless vital resources were withheld from the recovery programme 'we cannot prepare a defence programme independently of North America'.[13]

At the same time that Bevin accepted the preeminence of security considerations and abandoned attempts to achieve accommodation with the Soviet Union, the CoS were also persuaded to modify their approach to global strategy. From June 1947 the Middle East had been the geographical centre of their 'Commonwealth defence policy'. The defence of Western Europe was a much lower priority in defence planning. Even the Brussels Pact failed to convince two of the three Chiefs of Staff that a significant change was necessary. With the signing of the North Atlantic Treaty, however, which brought with it the prospect of American assistance in the defence of Western Europe, the CoS were prepared to modify their strategic plans.[14] In their 'Review of Defence Policy and Global Strategy' [DO(50)45] in May 1950 a much greater emphasis was given to fighting on the continent if war broke out.[15] At the same time less emphasis was placed on the defence of the Middle East (even though it remained an important consideration in British strategic planning). The CoS had also shifted their ground.

The signing of the North Atlantic Pact therefore reflected a consensus which had finally been reached between the views of the Foreign Office and those of the CoS. It was, however, a consensus which had been shaped by the growing security difficulties associated with the onset of the Cold War and a reluctance to participate in the moves towards a more supranational Europe. It was also a consensus which had only been reached after a period in which British foreign and defence policy had suffered from significant contradictions. These had arisen from an attempt to achieve an independent 'third power' role as leader of Western Europe and a broader Western Union, while concurrently pursuing a close, and necessarily dependent, relationship with the United States. The formation of the North Atlantic

Alliance also helped to overcome contradictions between the desire of the Foreign Office to play a leading role in coordinating the economic, political and military policies of Western European states and the CoS reluctance to provide military capabilities to perform such a role effectively. American support for Western European defence finally persuaded Britain to take the plunge and provide more permanent military support for the continent.

The consensus which had been arrived at focused not only on the North Atlantic Treaty but also on the broader global strategic framework for British foreign and defence policy. By the end of 1948 Britain was engaged in three sets of negotiations. From April onwards joint teams from Britain and the United States had been engaged in combined strategic planning in an attempt to devise parallel war plans in the event of an emergency in the following twelve months. High-level Anglo-American strategic planning had become a reality after years of frustration on the British side. In October and November also the Prime Ministers Commonwealth conference had set in motion a more urgent search for greater cooperation between the Commonwealth states in the defence field. The scale of cooperation remained limited but finally, after three years of halting progress, Commonwealth defence cooperation was beginning to take off.[16] At the same time (September to December 1948) crucial decisions were being made in the Washington talks which were to lead to the signing of the North Atlantic treaty.

Despite important adjustments within Britain's Commonwealth defence policy, important progress had been made in developing the strategic unity of the Commonwealth and Western Europe in close alliance with the United States, which had been a key objective of defence planners since the end of the war. To the CoS, NATO represented an important pillar in the global framework of security they were endeavouring to build. Equally, for Attlee and Bevin the Commonwealth, Western Europe and close ties with the United States provided an important framework to maintain Britain's status as a Great Power and to preserve international order which was a prerequisite for the recovery of Britain in the postwar world. As with the Chiefs of Staff, NATO was seen in the Foreign Office as an important part of a wider global vision based on close cooperation with the United States. Bevin told Acheson on 2 April 1949 that:

> In the Middle East there were 100 million Moslems – potentially a force not to be ignored in the world balance of power. Britain provided the 'best window' on this area and . . . rather than trying to create a joint military pact, Britain and the United States should adopt a common line in developing the resources, particularly oil, needed for its defence.[17]

He went on to develop the same theme on South and South East Asia, where 60 per cent of the population were also Muslim, and Russia had an obvious opening to undermine Western interests. Britain, he argued, 'could exercise influence through Pakistan but would need American help'.[18] He proposed to set up a conference on South East Asia in which Britain, the USA, Australia and New Zealand could cooperate for economic and political purposes as distinct from a military pact. This was designed to achieve a common front from Afghanistan to Indochina which would make it possible to contain the Russians, rehabilitate and stabilise the area and provide communications across the world.

By 1949, therefore, British foreign and defence policy were to a large extent in harmony. For both the Foreign Office and the CoS, NATO was of crucial significance. It provided a coherent focus for European security for the first time in the postwar period and an important pillar in the broader global strategy which Britain was trying to develop. In particular it set the scene for a close international defence partnership with the United States.

Far from being the culmination of a masterplan, NATO should be seen as a pragmatic response to rapidly changing events. Alan Bullock has argued that 'in the first eighteen months after 1945 it became clear that none of the schemes considered during the war for conducting foreign policy after it was viable. This was true of the wartime alliance of the Council of Foreign Ministers, of the Western Union and of the United Nations'.[19] Bullock argues that this lack of a framework for international relations was a source of great frustration to Bevin. In a world in which the major structures of international order had been destroyed, Bevin was forced to search for new arrangements which would reestablish stability and promote British interests. Lord Franks has argued that during 'his first two years at the Foreign Office (which at the time appeared lacking in positive achievement) Bevin did develop an overall picture of the changes taking place in the world, and had evolved two or three lines of policy which he then had the time to lay down and impress on the Foreign Office as the basis of British foreign policy'.[20] Even during this difficult period Bevin had a number of broad objectives which he sought to pursue. There was no blueprint but a series of alternative visions which he pursued simultaneously in his search for a stable international order and the reestablishment of British power and independence. Once he had determined, however, that an accommodation with the Soviet Union was not possible he worked ceaselessly to establish his vision of a Western Union backed by the United States. Despite some early disagreements over the Middle East he received firm backing from his Prime Minister, Clement Attlee, in the pursuit of these objectives. Attlee had great confidence in Bevin and left him free to conduct foreign policy without interference while he focused on domestic issues and the radical changes being introduced.

It is certainly true that Bevin's policies were not always consistent or indeed successful. His primary initial objective of reaching a satisfactory settlement with the Soviet Union was dashed as the Cold War gathered momentum. In the Middle East his hopes of creating a framework in which to develop a new relationship with the countries of the region came to nothing. His initial ideas of pursuing a somewhat independent role as leader of social democratic governments in Western Europe likewise had little success. Contradictions were also evident between foreign and defence policy, especially over the continental commitment, which he found difficult to resolve.

One of the main themes of British foreign policy in the period from 1945 to 1948 (which has been stressed in this book but often neglected by historians), was the search for an independent 'Third Force' role. It has been argued that Bevin's aim of coordinating the 'middle of the planet' in order to enhance British power and independence was abandoned in the second half of 1948 and early 1949 in favour of an Atlantic Alliance and a close security relationship with the United States. In the immediate postwar period Foreign Office officials realised that, given the weakness of Britain's economic position, the task of playing a leading role in Western Europe and maintaining a world role would require American economic and military assistance, at least in the short term. It was hoped that the United States would underwrite Bevin's initiatives until Britain had recovered from the temporary effects of the Second World War and could pursue a more independent role. The Foreign Secretary's plans, however, proved to be too ambitious, especially as the Cold War gathered momentum and Britain's economic position grew more precarious. Britain was not able to provide the resources to pursue a global strategy and at the same time back up its aspirations to play a leading role in Europe. Far from simply underwriting Bevin's 'grand design' it gradually became clear to the Foreign Office that a much more substantial American role would be necessary to match the power of the Soviet Union. As a result the 'Third Force' idea gave way to the search for an Atlantic Alliance and a 'special relationship' with the United States.

Although NATO was the result of a consistent British attempt to secure American assistance for Western European defence, it also owed a great deal to the collapse of previous policies, particularly the search for an independent role for Britain. As John Kent and John Young have argued, 'although the birth of NATO, in which Britain played a formative role, heralded a new postwar world order, this stemmed from the failure of

Bevin's schemes for the reassertion of British power rather than from a long term plan to produce an Atlantic Alliance'.[21] This suggests that the widely-held view of Ernest Bevin rescuing the Western world from the Soviet threat by painstakingly pursuing farsighted policies from 1945 onwards designed to establish an Atlantic Pact needs substantial revision.[22]

Whether Bevin should be criticised for his role in the formation of NATO is a matter of dispute. Some left-wing members of the Labour party at the time condemned the Foreign Secretary for failing to break with the past, for taking the side of the capitalist United States against socialist Russia and for fostering the damaging illusion that Britain was still a world power. In the latter case, it is claimed, this illusion set the pattern for the postwar period and played a significant part in deflecting Britain from its true destiny as part of Western Europe.[23]

To criticise Bevin for not breaking with the past, however, ignores the radical change in traditional policy implicit in the policies he pursued. NATO in particular represented a decisive break with one of the most important traditions in British foreign policy – for it involved a permanent alliance with North America and Western Europe – a commitment which British governments had always refused to consider in the past.[24] In his support for a continental commitment strategy Bevin also acknowledged the need to move away from the classical peripheral or maritime strategy which for centuries had been the hallmark of 'the British way of war'. Bevin therefore was certainly not afraid to break with tradition.

The contemporary left-wing criticism that Bevin took the side of the capitalist United States against socialist Russia is clearly true. It ignores the fact, however, that Bevin was prepared to stand up to the United States when necessary and, initially at least, believed that Britain would be able, once it recovered, to pursue a more independent line in international affairs.[25] He may have been misguided in this judgement but this was not an unreasonable objective given Britain's past history as a Great Power and her role as one of the Big Three victorious powers in the Second World War. Bevin realised that only a close relationship with the United States could help Britain recover from the dire economic circumstances it faced. No other country could provide Britain with the economic aid that was necessary. Given the beneficial effects of Marshall Aid in laying the foundations for Britain's postwar economic recovery it is difficult to argue that Bevin and his colleagues were wrong in their judgement.

It is also difficult to be overly critical of Bevin's approach to the Soviet Union. He was not prepared to be 'barged around' by the Soviet Union and he often took a tough line with Molotov in the negotiations which took place at the Foreign Ministers Conferences between 1945 and 1947. As a social democratic trade union leader he also had a deep-seated suspicion of Communism. Nevertheless, the evidence suggests that while Bevin was prepared to stand up for British interests he remained anxious to come to an agreement with the Soviet Union long after many of his Foreign Office officials had given up hope.[26] It was only after continuing Soviet intransigence (as he saw it) in numerous negotiations up to late 1947, and hostile Soviet policies in Eastern Europe blocked any hope of agreement, that Bevin finally turned to his alternative plan of developing a security framework for Western Europe and the Atlantic area.

Critics argue that the postwar era might have been different if only Bevin had pursued a socialist foreign policy. Alan Bullock sums up the nature of this policy in the following terms:

> It rejected the pursuit of national interest and power politics as the traditional ruling classes' conception of foreign policy, in favour of friendship and cooperation between peoples whose 'real' interests, it was argued, were the same in all countries. It rejected (in theory at least) loyalty to the nation state in favour of loyalty to the world community, and it rejected imperialism in favour of equality between peoples and the end of economic exploitation. Capitalism and war, indissolubly linked, were seen as the twin evils which Labour was committed to eradicate. From this followed strong suspicions of the foreign policy of any capitalist government . . . the task of the Labour party was not only to fight capitalism at home but its natural expression abroad, imperialism, power politics and war. From this in turn followed the rejection of force in settling international disputes; international solidarity with other working class movements, support for left-wing parties and governments in other countries and strong sympathy for the Soviet Union as the only socialist state in existence.[27]

However, these were not ideas that commended themselves to Ernest Bevin. He had spent the war as a member of a coalition government fighting for survival in a world dominated by power politics. His experience led him to believe that such socialist foreign policy views were irrelevant and dangerous in the real world where force and intimidation were all too common. In such an anarchical world Bevin believed states had no alternative but to pursue their national interests as best they could. The art of statesmanship was to create a stable international order which allowed democratic states to cooperate and prosper without fearing the kind of violence and intimidation which had characterised the 1930s. The lessons of appeasement, not surprisingly, were at the forefront of the minds of British leaders after the Second World War and had a powerful influence on their decisions to try to contain the Soviet Union. The analogy between Hitler and Stalin was probably false. It seems unlikely with hindsight that the Soviet Union had any intention of invading Western Europe. That was not Bevin's main concern, however. His primary objective in organising Western defence efforts was to provide the necessary self-confidence for Western European states who genuinely believed that they were threatened with subversion. Such self-confidence was necessary to allow them to rebuild their political, social and economic structures after the traumas and devastation of the war. This does not seem unreasonable given the circumstances which prevailed in the 1940s.

Whether a military alliance like NATO was necessary to achieve such self-confidence is a matter of legitimate debate. For a long time, as we have seen, George Kennan did not think so. He believed it would militarise the confrontation between East and West and institutionalise the antagonism between the two halves of Europe. The attempt to overcome that antagonism would then become even more difficult to achieve. In this he was proved correct. It was not until the mid-1980s and the arrival of Mikhail Gorbachev that the process of dismantling the military confrontation in Europe began in earnest. Even Kennan, however, finally conceded that the feelings of insecurity were very real and that NATO was an indispensable instrument in the recovery of Western Europe. The fact that public opinion overwhelmingly supported the Pact in all of the signatory states suggests that proponents of the alliance, like Bevin, were correct in their judgements.

There is a further criticism of Bevin which is worthy of consideration. This is the view that through the policies he pursued, including the establishment of NATO, the Labour Foreign Secretary continued and reinforced the illusion of Britain's great-power status. In so doing, it is argued, he deflected Britain from facing up to the realities of its diminished role in the world and delayed the entry into Europe. There appears to be some substance in this argument. By attempting to develop the strategic unity of the Commonwealth and Western Europe in close alliance with the United States Bevin and his colleagues did have a vision of a global security system in which Britain would play a key role. The result of this approach was to put an enormous strain on Britain's economic resources and led to overstretch in the nation's defence capabilities for much of the Cold War period which followed. A case can also be made that it helped to encourage the belief that Britain could stay aloof from the process of integration which was just beginning to take place in Western Europe.

Such an argument, however, depends heavily on hindsight. Looking back from the 1990s it seems obvious that Britain was in decline in the

1940s and was destined to join Europe. What this view ignores, however, is the fact that the scale of Britain's decline was far from clear in the immediate aftermath of the Second World War. Britain had played a crucial rule in the defeat of Germany and not surprisingly continued to think in great-power terms. In the immediate postwar period Bevin believed that Britain's decline was temporary. After a period Britain would be able to resume her rightful role as a major power. As the scale of Britain's economic plight dawned on the government, however, there was a clear recognition that dependence on the United States required a longer-term partnership between the two states. This was reinforced by the belief that the threat from the Soviet Union could not be met by British resources alone. In this sense the attempt to involve the United States in the defence of Western Europe and in support of British interests in the Middle East and South East Asia was itself part of a recognition that Britain's power had declined. By the late 1940s the illusions had partially gone and there was a realistic understanding that Britain could only defend its continuing far-flung interests through a close economic, political and military alliance with the United States. Although independence was being given to a number of Britain's Asian colonies it was clear that the Empire could not be disbanded overnight.

As far as Europe was concerned, Bevin found himself in something of a dilemma. He had a genuine interest in playing a leading role on the Continent. This was a role he tried to pursue in the period up to 1947. Increasingly, however, in 1948, he was faced with two difficulties. Firstly, there was the growing threat from the Soviet Union which Western Europe was incapable of dealing with without the support of the United States. And secondly, there was the continental movement towards supranationalism which Bevin had little time for. Both circumstances conspired to undermine the Foreign Secretary's 1946 'grand design' to achieve greater cooperation between Britain and the Western European states.

Despite the priority of the Atlantic over the European option, however, the continental commitment and NATO itself provided an important symbol of Britain's involvement in Europe for the future. Far from missing 'the European bus', as many have argued, Bevin established an unprecedented commitment on which his successors could build. He cannot therefore be criticised for the fact that subsequent governments and foreign secretaries did not join the European community until the 1970s.

After the uncritical assessments of Bevin of the 1970s and early 1980s a trend emerged in the historiography of postwar British foreign policy in the late 1980s and early 1990s which began to question Bevin's achievements.[28] Such critical studies are to be welcomed in helping to provide a more balanced interpretation of Bevin's period as Foreign Secretary. This study has tried to show that there were flaws and weaknesses in British policies which should not be ignored. Equally many of the criticisms levelled against Bevin at the time and with hindsight are wide of the mark.

Despite the qualifications, on balance, Bevin's role in the formation of NATO was a major achievement and not a failure of policy. The Foreign Secretary's great skill was in establishing a broad framework for the conduct of British diplomacy while retaining a pragmatic approach in dealing with specific issues. He pursued his long-term objectives of greater cooperation with Western Europe, the Commonwealth and the United States without a dogmatic commitment to particular forms of association. He constantly adapted to changing circumstances and when he became convinced that British interests were better served by an Atlantic framework he put all of his energies into achieving this aim. He was not afraid to change direction when the circumstances demanded it or to give up the pursuit of an independent role when it became unrealistic. Although there was at times a vagueness about his proposals, what mattered to Bevin was not the detail but the initiative. Concessions and compromises were constantly made in order to preserve the primary objective of defending British interests. In the pursuit of these aims, timing and patience were all-important. The negotiations which led to the Dunkirk Treaty, the proposals which followed the general failure of the Council of Foreign Ministers in December 1947, and the initiative in March 1948 in approaching Marshall when the tensions of the Cold War increased, all reflected the Foreign Secretary's ability to recognise the right moment to act.

Don Cook has argued that Bevin was 'a big man in the fullest sense of the word. If his learning was untutored, his grasp of issues was acute, his intelligence intuitive and far-ranging, his negotiating experience vast, his vision wide. He was a rocklike figure of certitude and common sense, and no man to be blown off course.'[29] During 1948 and early 1949, on numerous occasions when the US administration seemed to be back-tracking or 'dragging its feet', Bevin kept prodding the American government, refusing to accept that the domestic restraints precluded further progress in discussions on an Atlantic Pact. Similarly when irritations arose from time to time over what other delegations, as well as Britain, regarded as French insensitivity and intransigence, Bevin played a key role in persuading the French government to modify its policies for the benefit of the wider objectives which were being sought. While the wholly uncritical assessments of the Foreign Secretary's diplomacy undoubtedly need to be modified there is no doubt that Bevin's achievements were considerable.

Bevin's biographer has argued that:

> Throughout its history the alliance by which we have lived since the Second World War has undergone a number of metamorphoses. It may be that in the 1990s the time is ripe for another. But we should not fail to recognise that the Western Union of which Bevin first spoke in January 1948 has provided – with all its imperfections and limitations – a framework in which Britain and Western Europe have been able to enjoy an independence, a prosperity and a security, never of course, immune from crises or guaranteed to continue, but in practice lasting 35 years, a period longer than that which separates the Vienna settlement after the Napoleonic Wars from the Revolutions of 1848 and twice as long as that which followed the Treaty of Versailles. This is a future which would have seemed beyond the dreams of most people in Europe in the troubled years of 1945–50, and the part Bevin played in securing it, at the time when the whole pattern of international relations was in flux and for the first time the British found their power to influence it much reduced, was an achievement comparable with any of his predecessors.[30]

This seems a fitting assessment of Bevin's considerable achievement. It remains to be seen whether British foreign secretaries in the 1990s, faced with similar uncertainties, will be able to make the same kind of contribution to European security that Bevin made in the late 1940s.

The signing of the North Atlantic Treaty was not an inevitable product of the Cold War. There were numerous occasions, between the 'Pentagon Talks' in March 1947 and April 1949, when the attempts to secure some formal American contribution to the security of Western Europe could have failed. Given the traditional American reluctance to become involved in peacetime alliances, the chances of failure were very high. The fact that this did not happen and the United States joined Canada and Western European states in the NATO alliance is a tribute to the work of numerous statesmen and officials from different countries on both sides of the Atlantic. Bevin's role should not be exaggerated. He was not the sole architect of the Atlantic Pact. In this respect some of the 'depolarised' literature which focuses on Britain's (and Ernest Bevin's) achievements needs some revision. But equally it cannot be denied that Britain played a key role in the formation of NATO, and Bevin's own contribution should not be belittled. It is difficult to avoid the conclusion that the British Foreign Secretary demonstrated a combination of flexibility, pragmatism and patience in the conduct of diplomacy which made a major, if not a decisive, contribution to the formation of NATO.

Appendix 1

THE DUNKIRK TREATY, 4 MARCH 1947

His Majesty The King of Great Britain, Ireland and the British Dominions beyond the Seas, Emperor of India, and

The President of the French Republic,

Desiring to confirm in a Treaty of Alliance the cordial friendship and close association of interests between the United Kingdom and France;

Convinced that the conclusion of such a Treaty will facilitate the settlement in a spirit of mutual understanding of all questions arising between the two countries;

Resolved to co-operate closely with one another as well as with the other United Nations in preserving peace and resisting aggression, in accordance with the Charter of the United Nations and in particular with Articles 49, 51, 52, 53 and 107 thereof;

Determined to collaborate in measures of mutual assistance in the event of any renewal of German aggression, while considering most desirable the conclusion of a Treaty between all the Powers having responsibilities for action in relation to Germany with the object of preventing Germany from becoming again a menace to peace;

Having regard to the Treaties of Alliance and Mutual Assistance which they have respectively concluded with the Union of Soviet Socialist Republics;

Intending to strengthen the economic relations between the two countries to their mutual advantage and in the interests of general prosperity;

Have decided to conclude a Treaty with these objects and have appointed as their Plenipotentiaries:-

His Majesty The King of Great Britain, Ireland and the British Dominions beyond the Seas, Emperor of India:

For the United Kingdom of Great Britain and Northern Ireland,

The Right Honourable Ernest Bevin, M. P., His Majesty's Principal Secretary of State for Foreign Affairs and

The Right Honourable Alfred Duff Cooper, His Majesty's Ambassador Extraordinary and Plenipotentiary at Paris;

The President of the French Republic;

For the French Republic,

His Excellency Monsieur Georges Bidault, Minister for Foreign Affairs, and

His Excellency Monsieur René Massigli, Ambassador Extraordinary and Plenipotentiary of the French Republic in London;

who, having communicated their Full Powers, found in good and due form, have agreed as follows:-

Article I

Without prejudice to any arrangements that may be made, under any Treaty concluded between all the Powers having responsibility for action in relation to Germany under Article 107 of the Charter of the United Nations, for the purpose of preventing any infringements by Germany of her obligations with regard to disarmament and demilitarisation and generally of ensuring that Germany shall not again become a menace to peace, the High Contracting Parties will, in the event of any threat to the security of either of them arising from the adoption by Germany of a policy of aggression or from action by Germany designed to facilitate such a policy, take, after consulting with each other and where appropriate with the other Powers having responsibility for action in relation to Germany, such agreed action (which so long as the said Article 107 remains operative shall be action under the Article) as is best calculated to put an end to the threat.

Article II

Should either of the High Contracting Parties become again involved in hostilities with Germany,

either in consequence of an armed attack, within the meaning of Article 51 of the Charter of the United Nations, by Germany against that Party,

or as a result of agreed action taken against Germany under Article I of this Treaty,

or as a result of enforcement action taken against Germany by the United Nations Security Council,

the other High Contracting Party will at once give the High Contracting Party so involved in hostilities all the military and other support and assistance in his power.

Article III

In the event of either High Contracting Party being prejudiced by the failure of Germany to fulfil any obligation of an economic character imposed on her as a result of the Instrument of Surrender or arising out of any subsequent settlement, the High Contracting Parties will consult with each other and where appropriate with the other Powers having responsibility for action in relation to Germany, with a view to taking agreed action to deal with the situation.

Article IV

Bearing in mind the interests of the other members of the United Nations, the High Contracting Parties will by constant consultation on matters affecting their economic relations with each other take all possible steps to promote the prosperity and economic security of both countries and thus enable each of them to contribute more effectively to the economic and social objectives of the United Nations.

Article V

(1) Nothing in the present Treaty should be interpreted as derogating in any way from the obligations devolving upon the High Contracting Parties from the provisions of the Charter of the United Nations or from any special agreements concluded in virtue of Article 43 of the Charter.

(2) Neither of the High Contracting Parties will conclude any alliance or take part in any coalition directed against the other High Contracting Party; nor will they enter into any obligation inconsistent with the provisions of the present Treaty.

Article VI

(1) The present Treaty is subject to ratification and the instruments of ratification will be exchanged in London as soon as possible.

(2) It will come into force immediately on the exchange of the instruments of ratification and will remain in force for a period of fifty years.

(3) Unless either of the High Contracting Parties gives to the other notice in writing to terminate it at least one year before the expiration of this period, it will remain in force without any specified time limit, subject to the right of either of the High Contracting Parties to terminate it by giving to the other in writing a year's notice of his intention to do so.

IN WITNESS WHEREOF the above-mentioned Plenipotentiaries have signed the present Treaty and affixed thereto their seals.

DONE in Dunkirk the fourth day of March, 1947, in duplicate in English and French, both texts being equally authentic.

[L.S.] ERNEST BEVIN
[L.S.] DUFF COOPER
[L.S.] BIDAULT
[L.S.] R. MASSIGLI

Appendix 2

THE OVERALL STRATEGIC PLAN, MAY 1947
(DO(47))44
(Also CoS(47)102(0)) (Retained by the Cabinet Office)

FUTURE DEFENCE POLICY
Report by the Chief of Staff

OBJECT

The object of this paper is to set out the fundamental principles which should govern our Future Defence Policy and to arrive at a clear statement of the basic requirements of our Strategy, on which the shape and size of our armed forces can subsequently be planned.

We have accordingly arranged the paper in two parts:-

PART I. – Commonwealth Defence Policy. This Part concludes with a definition of the fundamentals of our Defence Policy.

PART II. – The Strategy of Commonwealth Defence. This Part begins with a statement of the basic requirements of our Strategy and concludes with a statement in general terms of the basic tasks of our armed forces and the principles which should govern their shape and size in order to fulfil this Strategy.

Any examination of Future Defence Policy inevitably gives rise to some consideration of long-term political and economic developments, since these affect our security quite as much as the shape and size of the armed forces. We have therefore attached at Annex our views on the political and economic objects which should be pursued in support of Commonwealth Defence Policy.

PART I. – COMMONWEALTH DEFENCE POLICY

International Relations

2. The fulfilment of the main object of the United Nations, the maintenance of world peace, depends on the ability and readiness of the Great Powers to keep the peace.

The supreme object of British policy is to prevent war, provided that this can be done without prejudicing our vital interests.

3. The United Kingdom, as the senior member of the British Common-

wealth and a Great Power, must be prepared at all times to fulfil her responsibilities not only to the United Nations but also to herself as a Great Power. To fulfil her obligations, she must achieve a strong and sound economy which will give her the ability to expand industry and the armed forces immediately on a war basis.

4. Because of the Veto, the United Nations Organisation provides no security against war between the Great Powers. In this situation, we believe that the only effective deterrent to a potential aggressor is tangible evidence of our intention and ability to withstand attack and to hit back immediately. No measure of disarmament should be accepted without adequate guarantees of security. Our aim must be to refashion our forces and our war potential to meet the needs of the future. We must remain strong enough to demonstrate our ability to withstand and our intention to counter aggression at any time.

5. Whether, therefore, we are acting in pursuit of national policy or in support of the United Nations, it is necessary to maintain British forces in peacetime to deter aggression which might lead to a major war and to defend our own interests. In support of these forces there must be reserves of essential resources.

Importance of Commonwealth Unity

6. The security of the United Kingdom is the keystone of Commonwealth defence. The United Kingdom contains 60 per cent of the white man-power and industrial capacity of the Commonwealth and the bulk of her scientific development. The Commonwealth gains its strength through the united front that it presents to the world. If the United Kingdom were to succumb, disintegration of the Commonwealth would inevitably follow, because the Dominions would not be strong enough to stand alone. Thus the defence of the United Kingdom is the vital concern, not only of the people of this country, but also of each separate member of the Commonwealth. Equally, the United Kingdom alone without the support of the Commonwealth would lose much of its effective influence and flexibility of power. It is, therefore, essential that the machinery for close and continuous co-operation in Defence matters between the Dominions and the United Kingdom should be constantly reviewed and brought to the highest efficiency.

Possible Threats to World Peace

7. Although we do not regard a future war as inevitable, we cannot yet be sure that all the Great Powers are determined to keep the peace. Until the general political atmosphere improves, we cannot, therefore, rule out

the possibility of war with Russia, either by actual aggression on her part or by a miscalculation of the extent to which she can pursue a policy of ideological and territorial expansion short of war with the Democratic Powers.

8. The issue which cannot be avoided is that our Defence Policy must at present be based on the possibility of war with Russia. We do not consider that Germany by herself will constitute a danger for many years, but should a resurgent Germany again become a menace, it would be possible to adjust our Defence Policy, if we have meanwhile prepared against a presently greater threat.

9. We are convinced that we can reduce the risk of war if from now onwards we and our potential allies show strength and a preparedness to use this strength if necessary. Subject to this, we believe that the likelihood of war in the next five years is small; that the risk will increase gradually in the following five years, and will increase more steeply thereafter as the rehabilitation of Russia gathers momentum.

Characteristics of Russia as a Potential Enemy

10. The power of Russia as a potential enemy rests on the following factors:-

(a) Her very great superiority in man-power.

(b) She has vast territory and great resources, is practically self-contained economically, and has the benefit of wide dispersion of industries and centres of population, all under strict Government control.

(c) Her present political organisation and the degree of control exercised over the people by the Government would effectively stifle any public clamour against war. In addition, the high standard of security achieved renders our collection of intelligence difficult and makes it the more likely that Russia will have the advantage of surprise at the outset.

(d) She makes full use of Communist Parties in other countries, both to achieve an advantageous position for herself before the outbreak of war and to undermine the effort of her enemies when war has broken out.

On the other hand, Russia suffers from the following elements of weakness:-

(e) Large parts of Russia have been completely devastated.

(f) The population is ignorant and ill-educated and the standard of living is low. She is at present inferior to the Western Powers both industrially and technically and though her industrial potential and technical ability are growing, she has still a long way to go.

(g) Her transportation system is vulnerable and comparatively undeveloped.
(h) Her oil production is barely sufficient for her needs and her main sources are badly placed strategically.

11. Against Russia as a potential enemy we must redress the balance in favour of the Commonwealth by:-
(a) Increasing and exploiting our present scientific lead. This applies particularly to the development of mass destruction weapons.
(b) Seeking to unite with us all powers which are determined to resist aggression.

European Allies

12. In the past we have relied on building up an alliance of European countries to unite with us from the very beginning in resisting aggression. There is now, however, no combination of European Powers capable of standing up to Russia on land, nor do we think that the probable military capabilities of an association of European States at present justify us in relying upon such an association for our defence.

Nevertheless, any time which we can gain to improve our defences would be of such value that every effort should be made to organise an association of Western European Powers, which would at least delay the enemy's advance across Europe.

Support from the United States

13. We must have the active and very early support of the United States. The United States alone, because of her man-power, industrial resources and her lead in the development of weapons of mass destruction, can turn the balance in favour of the Democracies. Apart from other considerations the United States will for some years at any rate, be the sole source from which we can draw a supply of atomic bombs.

Threat of Russian Expansion

14. The Russian policy of territorial and ideological expansion by the absorption of satellite States and by the spread of Communism in peace constantly threatens various countries whose continued integrity and independence profoundly affect Commonwealth security. Our interests are challenged, not only throughout Europe but also in the Middle East and throughout the world.

15. The area in which Russian expansion would be easiest and at the same time would hurt us most would be the Middle East. We may be sure that if we abandon our position there in peace Russia will fill the vacuum.

16. Our experience in other areas such as Eastern Europe has shown that when Russia gains control our economic interests are forfeited and our communications are cut. The first impact of Russian expansion into the Middle East would therefore be upon our oil supplies and upon Commonwealth sea and air communications. The importance to us of present and potential oil supplies in the area is as great, if not greater, than ever, particularly in peace. The importance of the Middle East as a centre of Commonwealth communications remains, and will remain, beyond question.

If the use of the Middle East communications was denied to us it would be necessary to divert our supplies round the Cape or across Central Africa, which would increase immeasurably the burden on our resources. Moreover, our strategic signal communications would be disrupted.

17. The powerful position which Russia would acquire by linking the Middle East countries to her influence and economy would prepare the way for further infiltration into both Asia and Africa. If Russia were to establish herself in the Middle East in peace or war, her power and influence would dominate the Moslem world and would be likely to spread eastwards through India, Burma and Malaya; southwards through the Sudan; and westwards in North Africa.

18. In all these areas cells of communism exist, but so far in isolation. Once Russia is established in the Middle East she will create from these isolated cells a comprehensive and unified organisation. This would seriously undermine our strategy and economic interests in all these areas. Her eastward expansion would threaten the security of India, our control of sea communications in the Indian Ocean and our resources of oil, tin and rubber. Her westward expansion would create a new threat to our Atlantic sea communications already likely to be gravely endangered.

19. Moreover, by ejecting the influence of the Western Powers from the Middle East, Russia would be securing her most vulnerable flank. It is from the Middle East area that her own vital oil industry and new industrial centres can most effectively be threatened. At the same time we should be placed in the position of having to be prepared to meet direct attacks on our own territories and interests in Africa, Aden, the Mediterranean and India and on our communications in the Indian Ocean.

20. To sum up, if Russia secured control of this area not only would we lose very important resources and facilities but she would acquire a position of such dominating strategic and economic power that it would be fatal to

our security. It is therefore vital that we must retain a firm hold on the Middle East. This can only be achieved by our physical presence there in peace and by tangible evidence of our intention to remain.

An important contribution to the security of our position will be the continued independence of Greece and Turkey.

21. The need to retain our strategic and economic position in the Middle East is of equal importance if we should be engaged in war with a power other than Russia. This is demonstrated by the fact that in two world wars we have had to defeat Germany in the Middle East.

Implications of New Weapons

22. The main implications of the new weapons likely to be available by the critical period about 1956 may be summarised as follows:-

(a) The possibility exists of achieving rapid and decisive results by the use of mass destruction weapons against economic key targets and the civil population.

(b) Owing to the vastly greater destructive power of atomic and biological weapons, acceptable standards of defence have gone up immeasurably. Within the next ten years there is little possibility that these higher standards of defence can be reached.

(c) There are greater possibilities than before of surprise attack, since the preparations required to deliver decisive attacks with the new weapons could be on a smaller scale than with conventional weapons. Militarily we must be prepared to exploit any such opportunity, although politically we are always likely to be severely handicapped.

(d) The potential threat to our sea communications will be greater than at any time in the last war.

Russian Technical Development

23. All our intelligence sources indicate that Russia is striving, with German help, to improve her military potential; and to catch up technically and scientifically.

We must expect that from 1956–57 Russia will probably be in a position to use some atomic bombs and biological warfare; that she may have developed, probably with German advice and technical assistance, rockets, pilotless aircraft, a strategic bomber force and a submarine force; and that she will continue to maintain very large land forces, a considerable proportion of which may be equipped and trained up to Western standards.

United Kingdom Intelligence Organisation

24. It is of the greatest importance that our Intelligence Organisation should be able to provide us with adequate and timely warning. The smaller the armed forces the greater is the need for developing our Intelligence Services in peace to enable them to fulfil this responsibility.

Vulnerability of the United Kingdom

25. The advent of mass destruction weapons and other new means of offence has greatly increased the vulnerability of the United Kingdom with her dense and concentrated population and industries. We do not think that it will be possible by purely defensive action to prevent the delivery of all weapons of mass destruction, and the effect of even a small number will be proportionately greater in the United Kingdom than in a large country with a widely dispersed industry.

26. In spite of our industrial and technical lead and in spite of the assistance of allies, including America, we should be unable to prevent the vastly superior land forces of Russia overrunning North-West Europe. From this position rockets and other long-range missiles might, if the build-up cannot be impeded, cause irreparable damage or even the elimination of the United Kingdom, even without recourse to atom attack. It is essential that before such destruction – from which we might never recover – could be achieved, we ourselves should assume the initiative and destroy the enemy's means of making war. The vulnerability of this country to modern weapons would bring the war to its climax much earlier than in the past. This initiative must be assumed from the outset. This entails not only the readiness of offensive forces, but the presence in this country of the essential resources to maintain them at intensive rates immediately on the outbreak of war.

Offensive Action

27. If we are to impede the enemy build-up in Western Europe and to strike at the enemy's means of making war we must possess Air Forces capable of penetrating into enemy territory. The time required to achieve decisive results with conventional weapons is open to doubt. It may be, therefore, that weapons of mass destruction will have to be used to achieve decisive results before irreparable damage to our industries has been caused. It is only by early offensive action that the weight of attack on the United Kingdom can be materially decreased.

Use of Mass Destruction Weapons

28. Thus to achieve victory or avoid defeat, it may be essential for us to use weapons of mass destruction.

On the other hand, in view of the small number of atom bombs and possibly of biological warfare and chemical warfare weapons required to knock out the United Kingdom, and the relatively small operational effort involved, the margin between victory and defeat, if once they are used, will be an extremely narrow one.

29. It may, therefore, be argued that, in view of the vulnerability of the United Kingdom, it would be to our advantage if they were not used by either side, and that for that reason we should favour their abolition in peace and not initiate their use in war. Nevertheless, their abolition by international convention would give us no guarantee of immunity, since their production in war, as a result of the development of atomic energy for other purposes, would be possible in a matter of months.

30. Furthermore, we are convinced that Russia's attitude towards the use of these weapons will be determined solely by self-interest. She will no doubt appreciate that a comparatively small number of bombs will knock us out, and that it will be impossible for us, by action taken solely against the aircraft delivering the weapons, to prevent this small number being delivered.

The only means whereby we can prevent her using them, therefore, is by facing her with the threat of large-scale damage from similar weapons if she should employ them. This threat can only be achieved by evidence of our ability to use weapons of mass destruction on a considerable scale from the outset.

In addition we believe that the knowledge that we possessed weapons of mass destruction and were prepared to use them would be a most effective deterrent to war itself.

31. The decision whether or not to use these weapons obviously cannot be taken now. The one certain point is that it must be a cardinal principle of our policy to be prepared, equipped and able to use them immediately.

Bases for Offensive Action

32. Irrespective of weapons used, bases for launching the air offensive are essential. There are three possible main areas from which effective offensive action can be launched against the USSR.

They are:-

(a) The United Kingdom. – The United Kingdom provides the best base for mounting an air offensive because of the airfields and resources that exist there. In this country, moreover, we can make full preparations in peace for the development of an immediate offensive in war and for the subsequent build-up for our allies.

The United Kingdom is, however, likely to be subject to a heavy scale of air attack and its capacity to sustain large offensive forces may therefore be limited; moreover, there are important areas of Russia which, for many years at any rate, will remain out of range of aircraft based in the United Kingdom.

We cannot, therefore, rely upon the United Kingdom as our sole offensive base. Moreover, without the use of overseas bases for offensive action it would not be possible to limit to any extent the weight of Russian effort against the United Kingdom.

(b) The Middle East. – From the Middle East it is possible to reach many vital areas of Russia which cannot effectively be dealt with by forces based in the United Kingdom. In particular, it is by far the best base for attack on Russia's oil production, one of the weak points in Russia's war potential.

(c) North-West India. – From bases in North-West India large areas of Russia can be covered, including the Siberian industrial areas which cannot effectively be dealt with from any other bases likely to be available to us.

FUNDAMENTALS OF OUR DEFENCE POLICY

33. From all the above factors we can now deduce the fundamentals of our Defence Policy:–

(a) The supreme object of British policy is to prevent war, provided that this can be done without prejudicing our vital interests. This entails support of the United Nations and ability to defend our own interests.

(b) The most likely and most formidable threat to our interests comes from Russia, especially from 1956 onwards, and it is against this worst case that we must be prepared, at the same time taking every possible step to prevent it.

(c) The most effective step towards preventing war is tangible evidence that we possess adequate forces and resources, that we are fully prepared and that we have the intention and ability to take immediate offensive action.

(d) Essential measures required in peace to give us a chance of survival and victory in the event of war are:–

   (i) Retaining at a high state of readiness properly balanced armed forces for immediate use on the outbreak of war, with the necessary reserves of resources to support them.

   (ii) Maintaining the united front of the British Commonwealth and doing everything possible to ensure that in the event of war we have the immediate and active support of all its members.

   (iii) Ensuring that we have the active and early support of the United States of America and of the Western European States.

(iv) Increasing and exploiting our present scientific and technical lead, especially in the development of weapons of mass destruction.

(v) Actively opposing the spread of Russian influence by adopting a firm attitude to further Russian territorial and ideological expansion, particularly in all areas of strategic value to the defence of the British Commonwealth.

(vi) Arresting by all possible means the deterioration that has already begun in our own position and prestige in the Middle East, and encouraging the continued independence of Greece and Turkey.

(vii) Maintaining our Intelligence Organisations at a high standard of efficiency.

(viii) Being prepared to take offensive air action from the outset since the war will rapidly reach a climax and the endurance of the United Kingdom cannot be guaranteed for any considerable period against attacks by modern weapons, still less by weapons of mass destruction. The best bases for this offensive action are United Kingdom, Middle East and if possible North-West India.

(ix) Being ourselves prepared, equipped and able to use weapons of mass destruction as a part of this offensive action.

PART II – THE STRATEGY OF COMMONWEALTH DEFENCE

34. Having arrived at conclusions on the fundamental principles which should govern our future Defence Policy, we proceed to examine further the strategy which would be required to implement that policy in war.

Our examination shows that in a future war, time will be an all-important factor. The days when we could afford to remain on the defensive, while gathering our great strength for the knock-out blow, ended with the advent of the first atom bomb. A far higher degree of preparedness in peace is now imperative if we are to survive the opening phases of another war – a preparedness which must enable us to hit back at the outset to defend our very existence. Moreover, in view of the speed with which we could be knocked out, it is vital that we possess the ability by ourselves to withstand and counter the initial onslaught. This entails stockpiling of reserves in peace-time.

The weight and tempo of this onslaught may, however, be beyond our power to bear alone for more than a short period. This places a new value on Dominion and Allied support, and calls for much more rapid assistance than of old, even if it is thereby limited.

35. If war should be forced upon us, it is obvious that our first consideration must be the defence of the United Kingdom, which is both the focus of Commonwealth strength and also its most vulnerable point. Besides the actual defence of these islands, we must also defend the resources on which the Commonwealth must draw to prosecute a major war, and preserve the means by which the Dominions and our allies can come to our aid so that with our united strength we can develop an eventual all-out offensive. The control of sea communications is essential to the achievement of these aims.

BASIC REQUIREMENTS OF OUR STRATEGY

36. It is now apparent that in pursuance of our Defence Policy the following are the basic requirements of our strategy:–

(a) The defence of the United Kingdom and its development as an offensive base.
(b) The control of essential sea communications.
(c) A firm hold in the Middle East and its development as an offensive base.

These three pillars of our strategy must stand together. The collapse of any one of them will bring down the whole structure of Commonwealth Strategy.

To them we would add a fourth, which though not essential would give a most desirable addition of strength:–

(d) The co-operation of India: the provision of the necessary assistance to ensure her security; and the development of an offensive base in North-West India.

We discuss below methods of attaining these strategic aims.

Defence of the United Kingdom

37. We must be prepared for a situation in which the Russians attempt to build-up rocket and air forces within easy striking range of the whole of the United Kingdom. The reduction of attack by enemy aircraft to acceptable proportions, if weapons of mass destruction are not used, is not a novel problem but we can see no way of combating the rocket once it has been launched. As we have shown, the capacity to develop an air offensive is an essential defence against attack by absolute weapons and against the build-up of rocket attack. The forces required for this would also be required in the battle for air supremacy to defeat normal methods of air attack. The possibility of attempts at invasion, particularly by air, cannot be ruled out. Finally, it must not be forgotten that the whole conduct of our defence will depend on the continued flow of our supplies, which must be transported by sea.

38. We must have, therefore:–

(a) Strong air defences, including strong and up-to-date anti-aircraft and civil defences.

(b) An effective bomber force.

(c) Naval control and air superiority over, on and under the waters surrounding these islands, and along our sea communications.

(d) Sufficient land forces for defence against invasion on a limited scale and against raids.

39. We have already pointed out that the security of the sea communications of the United Kingdom is an essential part of the defence of the United Kingdom both against invasion and to sustain the whole war effort of the country. This security must be extended worldwide to cover communications with the Dominions, with the United States and with sources of raw materials and essential supplies throughout the world.

The threat in Home waters and the North Atlantic would be immensely magnified by an enemy advance to the Channel ports and the Atlantic seaboard. Our powers of resistance to this threat would be much increased if we had the use of Irish bases.

40. We must also exercise control of sea communications to gain flexibility and mobility of deployment of all the Armed Forces wherever they may be required and to deny those advantages to the enemy. By control of the sea communications through the Mediterranean we could most quickly deploy our forces for the defence of the Middle East and obtain rapid assistance in that theatre from the United States and the Dominions, while at the same time by denying the use of that sea to the enemy, we confine him to difficult land communications and prevent him from obtaining a foothold in North Africa from which he might advance to outflank our defence of Egypt or to establish bases from which to threaten our Atlantic communications.

41. To enable us to exercise control of sea communications through the Mediterranean, we must retain our existing strategic possessions there and obtain additional base facilities on the North African coast. To these ends we must ensure that Spain does not fall under Russian domination, is prepared to resist aggression, and that the French North African dependencies are friendly to our cause.

42. To attain security and control of sea communications, naval and air forces, adequate and suitably organised to meet any threat or challenge, and bases for their effective operation will be required. At present it appears that the chief threats to our sea communications will be from fast submarine attack, air attack and minelaying. But the threat of surface attacks on our shipping must still be guarded against and the capacity of the potential

enemy to challenge our control of sea communications must be constantly watched and provided against.

Defence of the Middle East

43. The main problem of the defence of this area is the time factor and the effect of present political changes upon our position at the outbreak of war.

The vital strategic area of the Middle East is Egypt, since it possesses the essential air bases, ports, communications and manpower. Our defensive preparations therefore must be directed primarily to the retention of that area.

We cannot be certain of being able to defend our oil resources in the Middle East but we must make every effort to do so. It should be our aim to ensure that oil pipelines and other oil communications are as well placed strategically as possible. We should also endeavour, by all the means in our power, to develop sources of supply in less vulnerable areas in the Middle East and elsewhere and to build up our oil reserves, particularly by stock-piling in the United Kingdom and our other main base areas.

44. The defence of Egypt against a land attack from the north must be conducted in the area Southern Syria–Northern Palestine. The land forces which the Russians can deploy in this area would be operating at the end of long and difficult lines of communications, and would be reduced by maintenance difficulties. They would be further hampered if we were able to take early air action against their communications, from bases in the Middle East. A further advantage our Land Forces would enjoy would be the support of Naval Forces operating to their seaward flank. Provided we are established in the Middle East area before a Russian advance and provided early reinforcements can be obtained from the Dominions and from the United States it should be possible to defend our interests in the Middle East. The expenditure of resources required to recapture our position in the Middle East, if it is lost through our inability to concentrate there in time, will be out of all proportion to the expenditure needed to defend it if we are firmly established at the outbreak.

45. The problem is primarily one of time, i.e., whether we can get the necessary forces into position before the Russians can attack in strength. If the necessary arrangements are made for rapid assistance from the United States and the Dominions, we consider that an effective defence is well within our capacity provided that:-

(a) We have the right to re-enter Egypt on threat of war and develop the base facilities we shall require there.
(b) We have strategic rights in Palestine in peace.
(c) We retain the sovereignty of Cyprus.

(d) Turkey refuses Russian demands for strategic facilities in peace and opposes Russian invasion of her territory. This will modify the time factor to our advantage.

Continued independence of Greece will greatly encourage the Turks to stiffen their attitude.

Our position in the Middle East would be given greater depth if we obtain the Trusteeship of Cyrenaica.

East Africa has been suggested as a possible alternative to the retention of Middle East administrative facilities in peace. The time factor rules it out, however, as the cost in air and sea transport to meet the necessary speed of movement would be prohibitive. On the other hand East Africa would be a useful location for reserves.

Co-operation of India and her Defence

46. The importance of North-West India as an offensive base, together with the man-power and resources which India can provide, requires that every effort should be made to obtain an agreement with India whereby she will co-operate actively with us in war. We cannot, however, count upon this at present.

Except in the case of air defence, India should be able herself to undertake the main burden of defence. Should she agree to co-operate with us, we would have to provide the necessary assistance to ensure her security.

The same considerations would apply, in the case of a divided India, should a viable State emerge covering the north-west part of the country.

CONTROL-TASKS OF THE ARMED FORCES AND PRINCIPLES AFFECTING THEIR BUILD-UP

47. Our strategic needs lead to conclusions on the tasks of our armed forces which can now be stated in general terms. It is also possible to indicate certain general principles which should govern the nature and size of our future forces:–

(a) *Research and Development*

Our first requirement is to build up our organisation for scientific research and development to a level which will ensure that we can maintain our technical superiority on the one hand, and on the other provided the necessary information to enable us to decide in good time on our re-equipment policy.

(b) *Offensive Force*

As our ability to strike will represent both a very strong deterrent to aggression and one of our principal means of defence, the development of an air offensive force must be given high priority.

(c) *Defence of the United Kingdom*

The security of the United Kingdom is of vital importance. While the possession of a powerful offensive force is essential to our security, the development of our active and passive air defence organisations, in all their aspects, must be complementary to the build-up of our air striking force.

The Army must provide for the manning of the anti-aircraft defences; readiness to aid the Civil Power; and defence against invasion, primarily by air.

The Navy and the R. A. F. must ensure control over the waters surrounding these islands.

(d) *Control of Sea Communications*

While our ability to hit back and the knowledge that we possess a sound defence will be a very strong deterrent to a potential aggressor, once war starts the security of our sea communications will rapidly assume vital importance.

The task of the Navy, assisted by the R. A. F., will be to secure to our own use sea communications, not only in the approaches to the United Kingdom but world-wide with the Dominions, the United States and sources of supply, and also through the Mediterranean to the Middle East; at the same time denying them to the enemy.

The maintenance and development in peace of the necessary naval and air forces to ensure security and control against any threat or challenge will therefore continue to be of high importance.

At present it appears that the chief threats to our sea communications will be from fast submarine attack, air attack and minelaying. But the threat of surface attacks on our shipping must still be guarded against and the capacity of the potential enemy to challenge our control of sea communications must be constantly watched and provided against.

(e) *Defence of the Middle East and India*

The primary task of the Army, apart from the manning of anti-aircraft defences and readiness to aid civil power in the United Kingdom, will be to ensure the security of our Middle East base. Despite the possible risk of invasion of the United Kingdom by air we consider the provision of forces to meet our requirements in the Middle East must be given priority over the anti-invasion role in the United Kingdom.

Air forces will be required for the defence of the Middle East base and in support of the Army there. They will also be required in India if she is co-operating with us.

Naval forces capable of giving all necessary support to the Army's land battle will be required.

(f) *Combined Operations*

We do not foresee a necessity for major combined operations, in the form of assault landings by sea or air, in the early stages of a war. Nor would our military strength at the outset permit of their being undertaken. But it is impossible to forecast how the war would develop. Minor landings in furtherance of a land campaign already undertaken might be required. When the full strength of the Commonwealth and the Allies is built up, and an overseas operation is necessary, combined operations on a large scale may be required.

It will, therefore, be necessary in peace to provide for keeping the art of Combined Operations and Airborne Assault alive in all the Armed Forces, and for research, experiment and development in the technique required.

RECOMMENDATIONS

48. We recommend that the Defence Committee should:–

(a) Approve the fundamentals of our defence policy as given in paragraph 33 and accept the basic requirements of our strategy as given in paragraph 36 above.

(b) Approve the tasks of our armed forces as set out in paragraph 47 above as a basis for planning the shape and size of our future armed forces.

(c) Take note of our views on the relations between our Defence policy and National policy, as set out in the Annex.

(d) Direct that an examination be made of the desirability of increased dispersal of industry throughout the Commonwealth, as suggested in the Annex.

(Signed)

J. H. D. CUNNINGHAM
MONTGOMERY of ALAMEIN
W. F. DICKSON (V.C.A.S.)

Ministry of Defence, S.W.1,
22nd May, 1947

ANNEX

National Policy

1. Our examination of Future Defence Policy has led to the conclusion that long-term political and economic developments are factors as vital to our future security as any adjustments in the shape and size of the Armed Forces. We, therefore, give below our views on the relations between Defence Policy and both Foreign [and] Economic Policies.

2. We recognise that His Majesty's Government are committed to certain policies which may profoundly affect our Commonwealth strategic position, e.g., the reference of the Palestine question to the United Nations and the transfer of power in India. Nevertheless, we feel strongly that we should strive with all the means at our disposal, in consultation, as appropriate, with the Dominions and the United States, to achieve certain definite objects of policy.

3. Our examination shows that in support of Commonwealth defence policy the following objects should be pursued:–

- (a) We must support and strengthen the United Nations organisation, and seek to make it effective as a means of preventing war.
- (b) We should insist upon adequate guarantees of security before any measures of disarmament are undertaken.
- (c) We should continue to do everything possible to combat the spread of communism so as to prevent our position being prejudiced before the outbreak of war.
- (d) We should strengthen the links with the Dominions, including Eire, and all parts of the Colonial Empire.
- (e) We should have the closest possible tie-up with the United States.
- (f) We should strive for an agreement with India, or any part of India, which allows us all essential military facilities.
- (g) We must encourage the building up of a strong Western Region of Defence, with France as its key-stone, and ensure that Germany does not become a Russian Satellite.
- (h) We must strive to ensure that the solution of the Spanish problem is favourable to the democracies.
- (i) We should not adopt any policy in peace which might lead to difficulties in our obtaining in war our strategic requirements in the French or Spanish North African dependencies.
- (j) We must retain our essential strategic requirements in Palestine in peace.
- (k) We should negotiate a treaty with Egypt which will:–
  - (i) Safeguard our right of re-entry into that country on the threat of war;

(ii) Ensure the maintenance in peace of the minimum base facilities which we shall require on the outbreak of war.

(l) Our position in the Middle East would be given greater depth if we obtained military rights in Cyrenaica.

(m) We should not relinquish our sovereignty over Cyprus.

(n) We must not allow Russia to establish her influence in Libya or Somalia when the final deposition of these countries is determined.

(o) Nothing should be allowed to interfere with the improvement of our relations with the Arab States.

(p) We must ensure the integrity and independence of Greece and Turkey in peacetime and their capacity and willingness to resist aggression in war.

(q) We must support the continued independence of Persia and Afghanistan.

(r) We must encourage and resist the development of oil production in the less vulnerable areas of the Middle East and elsewhere.

(s) We must build up by stockpiling reserves of resources essential for our war-making capacity, particularly in the United Kingdom.

Economic Policy

The outstanding consideration from a strategical point of view is that the economy of the United Kingdom should be thoroughly sound and able to support a powerful war potential and adequate armed forces.

The greatly increased vulnerability of the United Kingdom suggests that dispersion of industry, not only within the United Kingdom, but also the less vulnerable Dominions, is of importance. Dispersion within the United Kingdom is already accepted as one of the considerations in our industrial planning. The practicability of a limited dispersion overseas of those industries vital to the prosecution of a war will depend on economic considerations. It is urgent that a thorough examination of this subject should be carried out by experts.

Appendix 3

THE BRUSSELS TREATY, 17 MARCH 1948

His Royal Highness the Prince Regent of Belgium, the President of the French Republic, President of the French Union, Her Royal Highness the Grand Duchess of Luxembourg, Her Majesty the Queen of the Netherlands and His Majesty the King of Great Britain, Ireland and the British Dominions beyond the Seas.

Resolved

To reaffirm their faith in fundamental human rights, in the dignity and worth of the human person and in the other ideals proclaimed in the Charter of the United Nations;

To fortify and preserve the principles of democracy, personal freedom and political liberty, the constitutional traditions and the rule of law, which are their common heritage;

To strengthen, with these aims in view, the economic, social and cultural ties by which they are already united;

To cooperate loyally and to coordinate their efforts to create in Western Europe a firm basis for European economic recovery;

To afford assistance to each other, in accordance with the Charter of the United Nations, in maintaining international peace and security and in resisting any policy of aggression;

To take such steps as may be held to be necessary in the event of a renewal by Germany of a policy of aggression;

To associate progressively in the pursuance of these aims other States inspired by the same ideals and animated by the like determination;

Desiring for these purposes to conclude a treaty for collaboration in economic, social and cultural matters and for collective self-defence;

Have appointed as their Plenipotentiaries:

His Royal Highness the Prince Regent of Belgium,

His Excellency Mr. Paul-Henri Spaak, Prime Minister, Minister of Foreign Affairs, and His Excellency Mr. Gaston Eyskens, Minister of Finance.

The President of the French Republic, President of the French Union,

His Excellency M. Georges Bidault, Minister of Foreign Affairs, and His Excellency Mr. Jean de Hauteclocque, Ambassador Extraordinary and Plenipotentiary of the French Republic in Brussels,

Her Royal Highness the Grand Duchess of Luxemburg,

His Excellency M. Joseph Bech, Minister of Foreign Affairs, and His

Excellency Mr. Robert Als, Envoy Extraordinary and Minister Plenipotentiary of Luxembourg in Brussels,

Her Majesty the Queen of the Netherlands,

His Excellency Baron C.G.W.H. van Boetzelaer van Oosterhout, Minister of Foreign Affairs, and His Excellency Baron Binnert Philip van Harinxma thoe Slooten, Ambassador Extraordinary and Plenipotentiary of the Netherlands in Brussels.

His Majesty the King of Great Britain, Ireland and the British Dominions beyond the Seas, for the United Kingdom of Great Britain and Northern Ireland,

The Right Honorable Ernest Bevin, Member of Parliament, Principal Secretary of State for Foreign Affairs, and His Excellency Sir George William Rendel, K.C.M.G. Ambassador Extraordinary and Plenipotentiary of His Britannic Majesty in Brussels, who, having exhibited their full powers found in good and due form, have agreed as follows:

Article I

Convinced of the close community of their interests and of the necessity of uniting in order to promote the economic recovery of Europe, the High Contracting Parties will so organize and coordinate their economic activities as to produce the best possible results, by the elimination of conflict in their economic policies, the coordination of production and the development of commercial exchanges.

The cooperation provided for in the preceding paragraph, which will be effected through the Consultative Council referred to in Article VII as well as through other bodies, shall not involve any duplication of, or prejudice to, the work of other economic organizations in which the High Contracting Parties are or may be represented but shall on the contrary assist the work of those organizations.

Article II

The High Contracting Parties will make every effort in common, both by direct consultation and in specialized agencies, to promote the attainment of a higher standard of living by their peoples and to develop on corresponding lines the social and other related services of their countries.

The High Contracting Parties will consult with the object of achieving the earliest possible application of recommendations of immediate practical interest, relating to social matters, adopted with their approval in the specialized agencies.

They will endeavour to conclude as soon as possible conventions with each other in the sphere of social security.

Article III

The High Contracting Parties will make every effort in common to lead their peoples towards a better understanding of the principles which form the basis of their common civilization and to promote cultural exchanges by conventions between themselves or by other means.

Article IV

If any of the High Contracting Parties should be the object of an armed attack in Europe, the other High Contracting Parties will, in accordance with the provisions of Article 51 of the Charter of the United Nations, afford the Party so attacked all the military and other aid and assistance in their power.

All measures taken as a result of the preceding Article shall be immediately reported to the Security Council. They shall be terminated as soon as the Security Council has taken the measures necessary to maintain or restore international peace and security.

The present Treaty does not prejudice in any way the obligations of the High Contracting Parties under the provisions of the Charter of the United Nations. It shall not be interpreted as affecting in any way the authority and responsibility of the Security Council under the Charter to take at any time such action as it deems necessary in order to maintain or restore international peace and security.

Article VI

The High Contracting Parties declare, each so far as he is concerned, that none of the international engagements now in force between him and any other of the High Contracting Parties or any third State is in conflict with the provisions of the present Treaty.

None of the High Contracting Parties will conclude any alliance or participating in any coalition directed against any other of the High Contracting Parties.

Article VII

For the purpose of consulting together on all the questions dealt with in the present Treaty, the High Contracting Parties will create a Consultative Council, which shall be so organized as to be able to exercise its functions continuously. The Council shall meet at such times as it shall deem fit.

At the request of any of the High Contracting Parties, the Council shall be immediately convened in order to permit the High Contracting Parties to consult with regard to any situation which may constitute a threat to peace, in whatever area this threat should arise; with regard to the attitude to be adopted and the steps to be taken in case of a renewal by Germany of an aggressive policy; or with regard to any situation constituting a danger to economic stability.

Article VIII

In pursuance of their determination to settle disputes only by peaceful means, the High Contracting Parties will apply to disputes between themselves the following provisions:

The High Contracting Parties will, while the present Treaty remains in force, settle all disputes falling within the scope of Article 36, paragraph 2, of the Statute of the International Court of Justice by referring them to the Court, subject only, in the case of each of them, to any reservation already made by that Party when accepting this clause for compulsory jurisdiction to the extent that that Party may maintain the reservation.

In addition, the High Contracting Parties will submit to conciliation all disputes outside the scope of Article 36, paragraph 2, of the Statute, of the International Court of Justice.

In the case of a mixed dispute involving both questions for which conciliation is appropriate and other questions for which judicial settlement is appropriate, any Party to the dispute shall have the right to insist that the judicial settlement of the legal questions shall precede conciliation.

The preceding provisions of this Article in no way affect the application of relevant provisions or agreements prescribing some other method of public settlement.

Article IX

The High Contracting Parties may, by agreement, invite any other State to accede to the present Treaty on condition to be agreed between them and the State so invited.

Any State so invited may become a Party to the Treaty by depositing an instrument of accession with the Belgian Government.

The Belgian Government will inform each of the High Contracting Parties of the deposit of each instrument of accession.

Article X

The present Treaty shall be ratified and the instruments of ratification shall be deposited as soon as possible with the Belgian Government.

It shall enter into force on the date of the deposit of the last instrument of ratification and shall thereafter remain in force for fifty years.

After the expiry of the period of fifty years, each of the High Contracting Parties shall have the right to cease to be a party thereto provided that he shall have previously given one year's notice of denunciation to the Belgian Government.

The Belgian Government shall inform the Governments of the other High Contracting Parties of the deposit of each instrument of ratification and of signed the present Treaty and have each notice of denunciation.

IN WITNESS WHEREOF, the above mentioned Plenipotentiaries have affixed thereto their seals.

DONE at Brussels, this seventeenth day of March 1948, in English and French, each text being equally authentic, in a single copy which shall remain deposited in the archives of the Belgian Government and of which certified copies shall be transmitted by that Government to each of the other signatories.

For Belgium:
(s) P.H. SPAAK; G. EYSKENS
For France:
(s) BIDAULT; J. DE HAUTECLOCQUE
For Luxembourg;
(s) JOS. BECH; ROBERT ALS
For the Netherlands:
(s) B. v. BOETZELAER; VAN HARINXMA THOE SLOOTEN
For the United Kingdom of Great Britain and Northern Ireland:
(s) ERNEST BEVIN; GEORGE RENDEL

Appendix 4

'THE PENTAGON PAPER', 1 APRIL 1948

FINAL DRAFT

The purpose of this paper is to recommend a course of action adequate to give effect to the declaration of March 17 by the President of support for the free nations of Europe. The recommendations made will require close consultation with political leaders of both parties in order that whatever policy is formulated may be a truly bipartisan American policy.

1. Diplomatic approaches to be made by the Government of the United States to the signatories of the Five-Power Treaty signed at Brussels on March 17, 1948 in order to secure their approval to its extension in the manner outlined below and to inform them of plans for the conclusion of a collective defense agreement for the North Atlantic Area, details of which are given below.

2. An immediate approach then to be made to Norway, Sweden, Denmark and Iceland, and (if the Italian elections are over) also to Italy, through diplomatic channels, by the United States, United Kingdom and France, with the consent of Benelux, with the object of explaining to them the scheme for a declaration by the President on the lines of that recommended in paragraph 3 below, and of ascertaining whether they would be prepared in such circumstances to accede to the Five-Power Treaty in the near future and to enter into negotiations for the North Atlantic Defense Agreement.

3. The President to announce that invitations had been issued to the United Kingdom, France, Canada, Norway, Sweden, Denmark, Iceland, The Netherlands, Belgium, Luxembourg, Eire, Italy, and Portugal (provided that secret inquiries had established the fact that these countries would be prepared to accept the invitations) to take part in a conference with a view to the conclusion of a collective Defense Agreement for the North Atlantic Area designed to give maximum effect, as between the parties, to the provisions of the United Nations Charter. In his statement the President would include a declaration of American intention, in the light of the obligations assumed by the signatories of the Five-Power Treaty and pending the conclusion of the Defense Agreement, to consider an armed attack in the North Atlantic Area against a signatory of the Five-Power Treaty as an armed attack against the United States to be dealt with by the United States in accordance with Article 51 of the United Nations Charter. The declaration would state that the United States would be disposed to

extend similar support to any other free democracy in Western Europe which acceded to the Five-Power Treaty. If, as a result of the inquiries referred to in paragraph 2 above, it appears that Norway, Sweden, Denmark, Iceland, and Italy, or any of them, do not wish to accede to the Five-Power Treaty at this stage, consideration would need to be given, in the light of the views of the above states, to the extension to them of some assurance of immediate support in case of an armed attack against them which they resisted resolutely. In any event, the declaration would be so phrased as to avoid inviting aggression against any other free country in Europe.

4. Simultaneously with this declaration an Anglo-American declaration to be made to the effect that the two countries are not prepared to countenance any attack on the political independence or territorial integrity of Greece, Turkey, or Iran, and that in the event of such an attack and pending the possible negotiations of some general Middle Eastern security system, they would feel bound fully to support these states under Article 51 of the Charter of the United Nations.

5. It is contemplated that the Defence Agreement referred to in paragraph 3 above would contain the following main provisions:

a. Preamble combining some of the features of the preambles to the Rio and Five-Power Treaties and making it clear that the main object of the instrument would be to preserve western civilization in the geographical area covered by the agreement. The Preamble should also refer to the desirability of the conclusion of further defense agreements under Article 51 of the Charter of the United Nations to the end that all free nations should eventually be covered by such agreements.

b. Provision that each Party shall regard any action in the area covered by the agreement, which it considers an armed attack against any other Party, as an armed attack against itself and that each Party accordingly undertakes to assist in meeting the attack in the exercise of the inherent right of individual or collective self-defense recognized by Article 51 of the Charter.

c. Provision following the lines of Article III, paragraph 2 of the Rio Treaty to the effect that, at the request of the State or States directly attacked, and until coordinated measures have been agreed upon, each one of the Parties shall determine the immediate measures which it will individually take in fulfillment of the obligation contained in the preceding paragraph and in accordance with the principle of mutual solidarity.

d. Provision to the effect that action taken under the agreement shall, as provided in Article 51 of the Charter, be promptly reported to the Security

Council and cease when the Security Council shall have taken the necessary steps to maintain or restore peace and security.

e. Delineation of the area covered by the agreement to include the continental territory in Europe or North America of any Party and the islands in the North Atlantic whether sovereign or belonging to any Party. (This would include Spitzbergen and other Norwegian Islands, Iceland, Greenland, Newfoundland and Alaska.)

f. Provision for consultation between all the Parties in the event of any Party considering that its territorial integrity or political independence is threatened by armed attack or indirect aggression in any part of the world.

g. Provision for the establishment of such agencies as may be necessary for effective implementation of the agreement including the working out of plans for prompt and effective action under b and c above.

h. Duration of ten years, with automatic renewal for five-year periods unless denounced.

6. When circumstances permit, Germany (or the three Western Zones), Austria (or the three Western Zones) and Spain should be invited to adhere to the Five-Power Treaty and to the Defense Agreement for the North Atlantic Area. This objective, which should not be publicly disclosed, could be provided for by a suitable accession clause in the Defense Agreement.

7. Political and military conversations to be initiated forthwith with the parties to the Five-Power Treaty with a view to coordinating their military and other efforts and strengthening their collective security.

Appendix 5

THE NORTH ATLANTIC TREATY

PREAMBLE

The Parties to this Treaty reaffirm their faith in the purposes and principles of the Charter of the United Nations and their desire to live in peace with all peoples and all governments.

They are determined to safeguard the freedom, common heritage and civilization of their people, founded on the principles of democracy, individual liberty and the rule of law.

They seek to promote stability and well-being in the North Atlantic area.

They are resolved to unite their efforts for collective defense and for the preservation of peace and security.

They therefore agree to this North Atlantic Treaty:

Article I

The Parties undertake, as set forth in the Charter of the United Nations, to settle any international disputes in which they may be involved by peaceful means in such a manner that international peace and security, and justice, are not endangered, and to refrain in their international relations from the threat or use of force in any manner inconsistent with the purposes of the United Nations.

Article 2

The Parties will contribute toward the further development of peaceful and friendly international relations by strengthening their free institutions, by bringing about a better understanding of the principles upon which these institutions are founded, and by promoting conditions of stability and well-being. They will seek to eliminate conflict in their international economic policies and will encourage economic collaboration between any or all of them.

Article 3

In order more effectively to achieve the objectives of this Treaty, the Parties, separately and jointly, by means of continuous and effective self-help and mutual aid, will maintain and develop their individual and collective capacity to resist armed attack.

Article 4

The Parties will consult together whenever, in the opinion of any of them, the territorial integrity, political independence or security of any of the Parties is threatened.

Article 5

The Parties agree that an armed attack against one or more of them in Europe or North America shall be considered an attack against them all; and consequently they agree that, if such an armed attack occurs, each of them, in exercise of the right of individual or collective self-defence recognized by Article 51 of the Charter of the United Nations, will assist the Party or Parties so attacked by taking forthwith, individually and in concert with the other Parties, such action as it deems necessary, including the use of armed force, to restore and maintain the security of the North Atlantic area.

Any such armed attack and all measures taken as a result thereof shall immediately be reported to the Security Council. Such measures shall be terminated when the Security Council has taken the measures necessary to restore and maintain international peace and security.

Article 6

For the purpose of Article 5 an armed attack on one or more of the Parties is deemed to include an armed attack on the territory of any of the Parties in Europe or North America, on the Algerian departments of France, on the occupation forces of any Party in Europe, on the islands under the jurisdiction of any Party in the North Atlantic area north of the Tropic of Cancer or on the vessels or aircraft in this area of any of the Parties.

Article 7

This Treaty does not affect, and shall not be interpreted as affecting, in any way the rights and obligations under the Charter of the Parties which are members of the United Nations, or the primary responsibility of the Security Council for the maintenance of international peace and security.

Article 8

Each Party declares that none of the international engagements now in force between it and any other of the Parties or any third state is in conflict with the provisions of this Treaty, and undertakes not to enter into any international engagement in conflict with this Treaty.

Article 9

The Parties hereby establish a council, on which each of them shall be represented, to consider matters concerning the implementation of this Treaty. The council shall be so organized as to be able to meet promptly at any time. The council shall set up such subsidiary bodies as may be necessary; in particular it shall establish immediately a defense committee which shall recommend measures for the implementation of Articles 3 and 5.

Article 10

The Parties may, by unanimous agreement, invite any other European state in a position to further the principles of this Treaty and to contribute to the security of the North Atlantic area to accede to this Treaty. Any state so invited may become a party to the Treaty by depositing its instrument of accession with the Government of the United States of America. The Government of the United States of America will inform each of the Parties of the deposit of each such instrument of accession.

Article 11

This Treaty shall be ratified and its provisions carried out by the Parties in accordance with their respective constitutional processes. The instruments of ratification shall be deposited as soon as possible with the Government of the United States of America, which will notify all the other signatories of each deposit. The Treaty shall enter into force between the states which have ratified it as soon as the ratification of the majority of the signatories, including the ratifications of Belgium, Canada, France, Luxembourg, the Netherlands, the United Kingdom and the United States, have been deposited and shall come into effect with respect to other states on the date of the deposit of their ratifications.

Article 12

After the Treaty has been in force for ten years, or at any time thereafter, the Parties shall, if any of them so requests, consult together for the purpose of reviewing the Treaty, having regard for the factors then affecting peace and security in the North Atlantic area, including the development of universal as well as regional arrangements under the Charter of the United Nations for the maintenance of international peace and security.

Article 13

After the Treaty has been in force for twenty years, any Party may cease to be a party one year after its notice of denunciation has been given to the Government of the United States of America, which will inform the Governments of the other Parties of the deposit of each notice of denunciation.

Article 14

This Treaty, of which the English and French texts are equally authentic, shall be deposited in the archives of the Government of the United States of America. Duly certified copies thereof will be transmitted by the Government to the Governments of the other signatories.

In witness whereof, the undersigned plenipotentiaries have signed this Treaty.

Done at Washington, the 4th day of April, 1949.

Sources

BRITAIN

1. Public Record Office (London)

CAB	Cabinet Office Papers
DEFE	Chiefs of Staff Committee Papers
PREM	Prime Minister's Office Papers
FO 371	Foreign Office General Files
FO 800	Bevin Papers

2. Private Papers

Alexander of Hillborough Papers (Churchill College, Cambridge)
Attlee Papers (Churchill College, Cambridge)
Dalton Papers (British Library of Political and Economic Science)

UNITED STATES

1. National Archives (Washington DC)

Record Group 59 General Records of the Department of State. Decimal Files

2. Harry S. Truman Library (Independence, Missouri)

Clark M. Clifford Papers
Harry S. Truman Papers
Henry L. Stimson Diaries
Oral Histories: Mathew J. Connelly; Loy E. Henderson; John D. Hickerson

3. George C. Marshall Library (Lexington, Virginia)

George C. Marshall Papers
Record Group 59: Policy Planning Staff Documents
Record Group 59: Records of Charles E. Bohlen

4. Library of Congress (Washington)

William D. Leahy Papers
Tom Connally Papers

CANADA

Public Archives of Canada (Ottawa)

Record Group 25
Record Group 2

Notes

INTRODUCTION

1. B. Zeeman, 'Britain and the Cold War: An Alternative Approach. The Treaty of Dunkirk Example', *European History Quarterly*, Vol. 16, no. 3, July 1986, p. 348.
2. W. H. McNeil, *America, Britain and Russia, Their Cooperation and Conflict 1941–46* (1953) and H. Seton-Watson, *Neither War nor Peace, The Struggle for Power in the Post-War World* (New York, 1959).
3. See W. Appleman Williams, *The Tragedy of American Diplomacy* (New York, 1962) and G. Alperowitz, *Atomic Diplomacy: Hiroshima and Potsdam* (New York, 1965).
4. See J. L. Gaddis, 'The Emerging Post-Revisionist Synthesis on the Origins of the Cold War', *Diplomatic History*, Vol. 7, no. 3, 1983.
5. See W. F. Kimball, 'Response to Gaddis', ibid. and B. Zeeman, 'Britain and the Cold War', op. cit.
6. W. F. Kimball, ibid., pp. 198–200.
7. Ibid.
8. B. Zeeman, 'Britain and the Cold War', op. cit., p. 344.
9. D. C. Watt, 'Rethinking the Cold War: A Letter to a British Historian', *The Political Quarterly*, Vol. 49, no. 4, 1978.
10. Ibid., p. 446.
11. See B. Zeeman, 'Britain and the Cold War', op. cit., p. 345.
12. Watt, 'Rethinking the Cold War', op. cit., pp. 455–6.
13. R. M. Hathaway, *Ambiguous Partnership: Britain and America, 1944–1947* (New York, 1981); T. H. Anderson, *The United States, Great Britain and the Cold War, 1944–1947* (Columbia, 1981); Cees Wiebes and Bert Zeeman, 'Her Verdrag van Duinkerken: Speelbal von Britze buitenlandse politiek, 1944–1947, *Inter-nationale Spectator*, Vol. 37, no. 8, 1983; N. Petersen 'Who Pulled Whom and How Much? Britain, the United States and the Making of the North Atlantic Treaty', *Millennium: Journal of International Studies*, Vol. 11, no. 2, 1982; E. Barker, *The British Between the Super Powers, 1945–1950* (London, 1983); J. Young, *Britain, France and the Unity of Western Europe, 1945–1951* (Leicester, 1984); W. R. Louis, *The British Empire in the Middle East, 1945–1951* (Oxford, 1984); W. C. Cromwell, 'The Marshall Plan, Britain and the Cold War', *Review of International Studies*, Vol. 8, no. 4, 1982; J. L. Gormly, 'The Washington Declaration and the "Poor Relations": Anglo-American Atomic Diplomacy, 1945–46', *Diplomatic History*, Vol. 8, no. 2, 1984; R. Ovendale, *The English-Speaking Alliance, Britain, the United States, the Dominions and the Cold War* (London, 1985); V. Rothwell, *Britain and the Cold War, 1941–47* (London, 1982); and S. Greenwood, 'Anglo-French Relations and the Origins of the Treaty of Dunkirk' (Doctoral dissertation, the University of London, 1982).
14. R. M. Hathaway, op. cit., p. 2.
15. B. Zeeman, 'Britain and the Cold War', op. cit., p. 346.

16. Ibid., pp. 346–7.
17. E. Reid, *Time of Fear and Hope: The Making of the North Atlantic Treaty, 1947–49* (Toronto, 1977).

1 WARTIME PLANNING FOR A POSTWAR EUROPEAN SECURITY GROUP, 1941–44

1. 'Defence Expenditure in Future Years: Interim Report by Minister for the Coordination of Defence', 15 December 1937. C. P. 316(37).
2. See M. Howard, *The Continental Commitment: The Dilemma of British Defence Policy in the Era of Two World Wars* (London, 1972) and B. Bond, *British Military Policy Between the Two World Wars* (Oxford, 1980).
3. 'European Appreciation' by the Chiefs of Staff Committee, 20 February 1939. DP (P) 44.
4. See O. Riste, 'Norway's "Atlantic Policy": The Genesis of North Atlantic Defence Cooperation', in A. de Staercke (ed.), *NATO's Anxious Birth: The Prophetic Vision of the 1940's* (London, 1985).
5. FO 371/29421, N 1307/87/30, 8 April 1941.
6. FO 371/29421, N 693/87/30, 12 March 1941; CAB 87/1, RP (41) 1, 24 February 1941.
7. FO 371/29422, N 6510/87/30, 14 November 1941.
8. See J. Lewis, *Changing Direction: British Military Planning for Post-War Strategic Defence, 1942–47* (London, 1988), pp. 4–16.
9. WM (42) 149th Concls (3), 3 November 1942, CAB 65/28.
10. See *The Memoirs of Lord Gladwyn* (London, 1972), p. 108.
11. Ibid., pp. 112–17.
12. Ibid., pp. 116–118.
13. FO 371/32525, U 742/742/70, 20 October 1942.
14. 'The United Nations Plan', CAB 66/63, WP (43) 31, 16 January 1943. See also Sir Llewellyn Woodward, *British Foreign Policy in the Second World War, Vol. V* (London, HMSO, 1976).
15. 'The Future of Germany', CAB 66/34, WP (43) 96, 8 March 1943.
16. Woodward, op. cit., p. 25.
17. Sir Orme Sargent was largely responsible for proposing the idea of confederations. See 'The Balkan Federation', FO 371/33134, R 3793/46/47.
18. Woodward, op. cit., p. 275–7.
19. For a discussion of Smuts's views see 'Cooperation in the British Commonwealth', memorandum by the Dominion Affairs Secretary, CAB 66/49, Secret DPM (44) 17, 7 April 1944.
20. See Woodward, op. cit., p. 183.
21. See *The Memoirs of Field Marshal Montgomery* (London, 1958), pp. 236–7.
22. 'Post-War Atlantic Bases', CAB 66/30, WP (42) 480, 22 October 1942; 'Norwegian Proposal for Post-war Anglo-American Norwegian Co-operation with regard to the Atlantic', FO 371/32832, N 5554/463/30.
23. 'The Future of Europe', CAB 66/48, WP (44) 181, 3 April 1944; 'The Future World Organizations: UN Plan', FO 371/40692, U 4102/180/70.
24. Woodward, op. cit. Spaak's representations came at the same time as a note

from Field Marshal Smuts urging the FO to prepare a paper on a Western European regional grouping for the Conference of Dominion Prime Ministers.

25. Duff Cooper, *Old Men Forget* (London, 1954), pp. 344–7.
26. Ibid., pp. 346–7.
27. Anthony Eden, *The Eden Memoirs: The Reckoning* (London, 1965), pp. 444–6.
28. Cooper, op. cit., p. 347.
29. Eden, op. cit., p. 444.
30. Ibid.
31. Ibid.
32. Ibid., p. 445.
33. Ibid.
34. Woodward, op. cit., p. 187.
35. The Americans had a fairly clear idea of Britain's policy towards a Western European group at this time. See 'British Plan for a Western European Bloc', Department of State (DoS) Files 840 OD/1–1645.
36. Woodward, op. cit., p. 188.
37. 'British Policy towards Western Europe', CAB 80/44, Secret CoS (44) 113, 3 June 1944.
38. Woodward, op. cit., p. 188.
39. After the Dunkirk Treaty in 1947 with France, Britain initially planned to use the bilateral treaty as the model for its approach to the Benelux countries. It was the Benelux countries themselves who argued for a wider multilateral arrangement. See pp. 69–70.
40. This was a continuous theme throughout much of the wartime planning. See T. A. Imogibhe, 'War-time Influences affecting Britain's Attitude to a Post-war Continental Commitment' (Doctoral thesis, University of Wales, 1975), pp. 151–77.
41. Woodward, op. cit., p. 188.

2 EMERGING DIFFERENCES BETWEEN THE CHIEFS OF STAFF AND THE FOREIGN OFFICE, 1944–45

1. See Lewis, op. cit., Chapter 3.
2. Ibid., p. 71.
3. CAB 81/40, PHP (44) 3rd Mtg (1), 27 January 1944; CAB 79/70, CoS (44) 50th Mtg (0) (3), 17 February 1944.
4. Ibid.
5. Lewis, op. cit., p. 71.
6. FO 371/40741A, U 6283/748/70, PHP (44) 17 (0) (Draft), 7 July 1944.
7. Ibid.
8. Ibid.
9. Ibid.
10. Ibid.
11. FO 371/40741A, U 6283/748/70, 12 July 1944.

12. Davison letter to Redman of 27 July 1944 in CAB 122/1566. Quoted in J. Lewis, op. cit., p. 119.
13. FO 371/40741 A, U 6793/748/70, 26 July 1944; CAB 21/1614, CoS (44) 248 Mtg (0) (14), 26 July 1944.
14. Ibid.
15. FO 371/40741 A, U 6793/748/70, CoS Sec Min 1289/4 27 July 1944; CAB 79/78, CoS (44) 249th Mtg (0) (7), 27 July 1944.
16. FO 371/40741 A, U 6793/748/70, 28 July 1944.
17. A. Bryant, *Triumph in the West* (London, 1959), p. 242.
18. CAB 66/53, WP (44) 409, 25 July 1944.
19. Ibid.
20. Lewis, op. cit., p. 121.
21. Ibid., p. 122.
22. FO 371/39080, C 11955/146/18, 28 August 1944.
23. CAB 79/80, CoS (44) 303rd Mtg (0) (5), 9 September 1944; CAB 80/87, CoS (44) 822 (0), 9 September 1944.
24. Ibid.
25. Ibid.
26. Ibid.
27. Woodward, op. cit., p. 205.
28. Ibid. Between August and September 1944 several representations were made by Sir Orme Sargent and Wilson, head of the Northern Department, to Eden complaining of 'careless talk' by the military authorities of the possibility of Anglo-Soviet hostility after the war. They feared this might come to the attention of the Americans and the Russians.
29. FO 371/39080, C 13518/146/18, 2 October 1944.
30. FO 371/39080, C 13518/146/18, 4 October 1944.
31. CAB 79/81, CoS (44) 323rd Mtg (0) (6), 2 October 1944.
32. FO 371/39080, C 13518/146/18, 4 October 1944.
33. Ibid.
34. Lewis, op. cit., p. 133.
35. CAB 81/95, PHP (44) 27 (0), 9 November 1944.
36. Lewis, op. cit., p. 122.
37. See C. Wiebes and B. Zeeman, 'Baylis on Post-war Planning' and J. Baylis 'A Reply' in *Review of International Studies*, Vol. 10, no. 3, 1984.
38. CAB 81/95, *op. cit.*
39. Ibid.
40. The report was determined to stress, however, that no move should be made to bring any part or all of Germany into a bloc against the Soviet Union until relations with her had irrevocably broken down. See Rothwell, op. cit., pp. 119–20.
41. FO 371/40471 B, U 8181/748/70, 21 November, 1944.
42. Lewis, op. cit., p. 163.
43. Ibid., p. 164.
44. Ibid., p. 165.
45. Ibid.
46. Ibid., p. 171.
47. Ibid., p. 169.
48. Ibid., p. 174.

49. Ibid., p. 176.
50. FO 371/40720, U 7917/8652/180/70, 5 October, 1944.
51. See P. A. Reynolds and E. J. Hughes, *The Historian as Diplomat: Charles Kingsley Webster and the United Nations, 1939–1946* (London, 1976), pp. 11, 20–1, 29, 30–4.
52. Woodward, op. cit., pp. 193–4.
53. Rothwell, op. cit., pp. 406–10.
54. Ibid.
55. Ibid.
56. Woodward, op. cit., pp. 193–4.
57. Ibid., pp. 194–6.
58. R. Ovendale, *The English-Speaking Alliance: Britain, the United States, the Dominions and the Cold War 1945–51* (London, 1985), p. 6.
59. Ibid.
60. E. Barker, *The British between the Superpowers, 1945–50* (London, 1983), p. 4.
61. V. Rothwell, op. cit., pp. 128–9.
62. R. Ovendale, op. cit., p. 9.
63. V. Rothwell, op. cit., pp. 136–9. See also *Foreign Relations of the United States (FRUS)*, The Conference at Yalta, (Washington, DC, 1955) and CAB 65/51, WM (45) 22, 19 February 1945.
64. CAB 65/53, WM (45) 39, 3 April 1945. See also R. Ovendale, op. cit., pp. 11–12.
65. C. M. W. Moran, *Winston Churchill: The Struggle for Survival, 1940–1965* (London, 1966), pp. 279–84.
66. V. Rothwell, op. cit., pp. 114–23.

3 POSTWAR ATTITUDES TOWARDS THE SOVIET UNION

1. A. Bullock, *Ernest Bevin: Foreign Secretary, 1945–1951* (London, 1983), p. 21.
2. D. Cook, *Forging the Alliance: NATO, 1945–1950* (London, 1989), p. 19.
3. FO 371/50912, 11 July 1945. See also R. Butler and M. E. Pelly (eds), *Documents on British Policy Overseas*, Series 1, Vol. 1 (London, 1984), no. 102.
4. Ibid.
5. See A. Adamthwaite, 'Britain and the World, 1945–1949: the View from the Foreign Office', in J. Becker and F. Knipping (ed.), *Power in Europe? Great Britain, France, Italy and Germany in a Post-war World, 1945–1950* (New York, 1986), pp. 13–14.
6. FO 50912/5471, op. cit.
7. FO 371/50919, 1 October 1945.
8. FO 371/50918, 1 October 1945.
9. A convergence in military and diplomatic thinking had been reached on this issue by the end of 1945. CAB 80/99, CoS (46) 43 (0), 13 February 1946.
10. P. Dixon, *Double Diploma: the Life of Sir Pierson Dixon* (London, 1968), pp. 194–9.

11. FO 371/50286, 8 October 1945.
12. FO 371/50920, 18 October, 1945.
13. FO 371/50919, Minutes by Ward, 12–19 October 1945.
14. FO 371/50920, 18 October 1945.
15. V. Rothwell, op. cit., p. 246.
16. Ibid.
17. FO 371/56763, 14 March, 1946.
18. FO 371/56830, 31 December 1945 and FO 371/52327, 16 January 1946. See also Rothwell, op. cit., pp. 247–50 and R. Ovendale, op. cit., pp. 29–54.
19. Ibid.
20. FO 371/56763, 14 March 1946.
21. Rothwell, op. cit., p. 251.
22. FO 371/56780, 23 February 1946.
23. FO 371/55581, 2 April 1946.
24. Rothwell, op. cit., p. 259.
25. FO 371/56908, 19 July 1946.
26. Rothwell, op. cit., p. 260.
27. Ibid.
28. A. Foster, 'The British Press and the Coming of the Cold War', in A. Deighton (ed.), *Britain and the First Cold War* (London, 1990).
29. FO 800/451, Def/46/3, 13 February 1946. This does not, however, mean that Bevin was committed to a long-term, exclusive Anglo-American relationship at this stage.
30. Rothwell, op. cit., p. 262. See also D. Cook, op cit., p. 55.
31. FO 371/66279, 28 May 1946.
32. FO 371/55593, 30 September 1946.
33. Rothwell, op. cit., pp. 268–70.
34. FO 371/66546, January 1947.
35. Ibid.
36. FO 371/66363, 12 February 1947.
37. Jebb to Dixon, 10 April 1947, folder of miscellaneous papers, 1947, *Dixon Papers*. See also Rothwell, op. cit., p. 274.
38. FO 371/67579, 21 and 29 May 1947.
39. A. Bullock, *Ernest Bevin: Foreign Secretary, 1945–51* (London, 1983), p. 52. See also K. O. Morgan, *The People's Peace: British History 1945–1989* (Oxford, 1991).
40. H. Dalton, *High Tide and After, Memoirs 1945–1960* (London, 1962), pp. 22–3.

4 TOWARDS A TREATY WITH FRANCE

1. FO 371/45551, E 6051/8/89, 5 August 1945.
2. See S. Greenwood, 'Ernest Bevin, France and "Western Union": August 1945–February 1946', *European History Quarterly*, Vol. 14, 1984, p. 322.
3. FO 371/50655, U 120/1/70, 9 January 1945.
4. Woodward, op. cit., pp. 329–38.
5. De Gaulle, *Memoires de Guerre, III: Le Salut, 1944–46* (Paris, 1959), p. 194.

6. FO 371/49069, Z 9595/13/17, 13 August 1945.
7. Ibid.
8. See J. Young, *Britain, France and the Unity of Europe, 1945–51* (Leicester, 1984), p. 14.
9. Ibid.
10. See D. Cook, op. cit., p. 21.
11. *Halifax Microfilm* (Churchill College, Cambridge) 410 4.11, Halifax to Churchill, 3 August 1945.
12. Ibid., 410 4.16 Cadogan to Halifax, 30 December 1945.
13. CAB 128/3, CM (45) 35, 25 September 1945.
14. See Greenwood, 'Bevin and "Western Union", 1945–6', op. cit., p. 327.
15. Ibid., p. 328.
16. Ibid.
17. FO 371/45731, UE 3689/3683/53, 17 August 1945.
18. See S. Greenwood, 'Bevin and "Western Union", 1945–6', p. 330.
19. Ibid., p. 325.
20. FO 371/50826, U 8136/445/70, October 1945.
21. FO 371/59957, Z 754/21/17, undated minute.
22. Greenwood, 'Bevin and "Western Union", 1945–46', op. cit., p. 332.
23. FO 371/59957, Z 909/21/17, 30 January 1946.
24. Greenwood, 'Bevin and "Western Union", 1945–46', op. cit., p. 333.
25. FO 371/58179, UR 5561/17/851, 27 March 1946; CAB 129/8, CP (46) 139, 15 April 1946.
26. Ibid.
27. Ibid.
28. FO 371/59958, Z 1406/21/17, 12 February 1946.
29. FO 371/59911, Z 259/120/72, 9 January 1946.
30. FO 371/59911, Z 2410/120/72, 13 March 1946.
31. FO 371/59911, Z 3319/0/17, 8 April 1946.
32. Ibid.
33. Ibid.
34. Ibid.
35. FO 371/58853, Z 3744/20/17, 18 April 1946.
36. FO 371/59953, Z 34–5/20/17, 11 April 1946.
37. See J. Baylis, 'Britain and the Dunkirk Treaty: The Origins of NATO', *The Journal of Strategic Studies*, Vol. 5, no. 2, 1092, p. 240.
38. FO 371/59911, Z 2410/120/72, 13 March 1946.
39. Ibid.
40. FO 371/22134, R 3793/46/67. Sir Orme Sargent had proposed the idea of various confederations in 1943.
41. FO 371/59911, Z 10754, 31 December 1946. Those attending the meeting included Mr Butler, Mr Warner, Mr Dening, Mr Troutbeck, Mr Hankey, Mr Mason, Mr Hayter, Mr Burrows and Sir A. Rumbold.
42. Especially as a result of the problems which had arisen over Iran, the Dardanelles as well as Germany and Soviet policies in Eastern Europe and the difficulties at the Foreign Ministers Conferences in London, Moscow, Paris and New York.
43. FO 371/59911, Z 10754, 31 December 1946.
44. Ibid.

45. Duff Cooper, op. cit., pp. 344–7.
46. Duff Cooper was not the only one to advocate that Britain should stand between the United States and the Soviet Union. Bevin himself seems to have hoped that Britain could play an intermediary role. See FO 371/59911, Z 10754, 31 December 1946.
47. Ibid.
48. FO 371/67670, Z 25/25/17, 31 December 1946.
49. Ibid.
50. Even this section in December 1947 was designed to bind the states of Western Europe through bilateral treaties, not through the kind of multilateral pact which eventually emerged in March 1948.
51. FO 371/67670, Z 291/25/17, 1 January 1947.
52. Ibid.
53. CAB 128/9, CM (47) 2, 6 January 1947.
54. See S. Greenwood, 'Return to Dunkirk: The Origins of the Anglo-French Treaty of March 1947', *The Journal of Strategic Studies*, Vol. 6, 1989, p. 40.
55. Whether opportunism or high politics was more important in the decision to conclude the Dunkirk Treaty is difficult to say. Greenwood tends to stress the former while Young in his book *Britain, France and the Unity of Europe, 1945–51*, op. cit., tends to put slightly more emphasis on the 'grand design'.
56. FO 371/67670, Z 965/25/17, 18 January 1947.
57. FO 371/67671, Z 1649/25/17, 7 February, 1947.
58. FO 371/67670, Z 965/25/17, 23 January 1947.
59. Ibid.
60. FO 371/67671, Z 1662/25/17, 15 February 1947.
61. Ibid. Sargent in particular thought that it should be sought regardless of the impact on the United States. See Rothwell, op. cit., p. 435.
62. FO 371/67671, Z 1931/25/17, 22 February 1947.
63. This was contained in a new Article One to the Treaty.
64. They both felt that complications might result if the signing was left until after the Moscow Conference.
65. Bevin would have preferred to have signed the Treaty at Calais, but Bidault argued in favour of the symbolism of Dunkirk. Dunkirk was where Anglo-French relations had ended in the war and he felt that the new start represented by the Treaty should be made from the same place.
66. Duff Cooper was disappointed over the failure to start Staff Talks after the Treaty was signed.
67. This is reflected, as we have seen, in the Post-Hostilities Planning Staff Report in November 1944.
68. Supporters of both forms of cooperation argued their cases in the Foreign Affairs debates in the House of Commons during 1946. Churchill put his views across in favour of a close military alliance with the United States in his Fulton Speech in March 1946.
69. See the Debates on Foreign Affairs in *Hansard* on 20–1 February, 4–5 June, 22–3 October and 12–14 November 1946.
70. Young op. cit., p. 50.
71. FO 371/73051, Z 1999/273/72/9, 9 March 1948.
72. For differing interpretations of the significance of the Dunkirk Treaty see

B. Zeeman, 'Britain and the Cold War: An Alternative Approach. The Treaty of Dunkirk Example', op. cit.

73. See *Hansard*, Vol. 416, Col. 762, 23 November 1945.
74. *Foreign Relations of the United States* (*FRUS*), 1945 (11), 629, 17 December 1945.
75. A. Adamthwaite, 'Britain and the World, 1945–9: the view from the Foreign Office', *International Affairs*, Vol. 61, 1985, p. 228.
76. See p. 55.
77. See note 61.

5 THE WESTERN UNION AND THE BRUSSELS PACT

1. CAB 131/1, DO (46) 1st Mtg, 11 January 1946 and CAB 13/1, DO (46), 22nd Mtg, 19 July 1946.
2. FO 800/451, Def 146/3, 13 February 1946.
3. CAB 128/4, CM (45) 50th Concls CA, 6 November 1945.
4. See E. Barker, op. cit., pp. 79–84.
5. FO 371/73045, Z 322/273/72/G, 7 May 1947.
6. FO 371/73045, Z 322/273/72/G, 3 June, 1947.
7. Ibid.
8. *FRUS*, Vol. 3, 1947, pp. 230–2.
9. This was a phrase used by Bevin in a speech to the National Press Club in Washington on 1 April 1949.
10. This was a phrase used by Dalton to the Cabinet on 27 March 1947.
11. See A. Bullock, *Ernest Bevin: Foreign Secretary: 1945–51* (London, 1983), p. 490.
12. FO 371/67673, Z 8579/251/17/G, 26 September 1947.
13. See E. Reid, *Time of Fear and Hope: The Making of the North Atlantic Treaty, 1947–1949* (Toronto, 1977), p. 33.
14. FO 371/67674, Z 11010/25/17/G, 17 December 1947.
15. Ibid.
16. See N. Henderson, *The Birth of NATO* (London, 1982), p. 1.
17. See Reid, op. cit., p. 37. See also D. Cook,, op. cit., p. 116.
18. FO 371/73045, Z 809/273/72/G. Sir Roger Makins asked for clarification on 21 January 1948.
19. FO 371/73045, Z 273/2173/72/G, 13 January 1948.
20. See Henderson, op. cit., p. 3.
21. CAB 129/23 CP (48) 6.
22. See *Foreign Relations of the United States*, Vol. 3 (Washington, 1974), pp. 4–6. See also John Kent and John Young, 'The "Western Union" Concept and British Defence Planning 1947–48', in R. Aldrich (ed.), *British Intelligence and Security Policy* (London, 1991). I am particularly grateful to John Kent and John Young for helping to establish for me the importance of the 'third force' idea.
23. FO 371/73045, Z 273/273/72/G, 16 January 1948.
24. See Henderson, op. cit., p. 6.
25. See Bullock, op. cit., p. 519.

26. Ibid., p. 520.
27. Ibid.
28. *The Economist*, 31 January 1948.
29. Ibid.
30. FO 371/73045, Z 554/273/72/G, 20 January 1948.
31. Ibid., 19 January 1948.
32. FO 371/73045, Z 561/273/72/G, 21 January 1948.
33. Ibid.
34. Ibid.
35. Ibid., 26 January 1948.
36. FO 371/73046, Z 937/273/72/G, 5 February 1948.
37. FO 371/73046, Z 896/273/72/G, 2 February 1948.
38. FO 371/73046, Z 937/273/72/G, 2 February 1948.
39. FO 371/73046, Z 1060/273/72/G, 7 February 1948.
40. FO 371/73045, Z 561/273/72/G, 26 January 1948. See also R. Ovendale, 'Britain, the USA and the European Cold War, 1945–48', *History*, June 1982.
41. FO 371/73046, Z 814/273/72/G, 28 January 1948.
42. FO 371/73045, Z 562/273/72/G.
43. FO 371/73045, Z 814/273/72/G.
44. Ibid.
45. FO 371/73047, Z 1308/273/72/G, 13 February 1948.
46. Ibid.
47. Ibid.
48. FO 371/73048, Z 1400/273/72/G, 16 February 1948.
49. Ibid. The revisionist literature includes: W. A. Williams, *The Tragedy of American Diplomacy* (New York, 1962); D. F. Fleming,, *The Cold War and its Origins* (New York, 1961); D. Horowitz, *The Free World Colossus: A Critique of American Foreign Policy in the Cold War* (New York, 1971); G. Alperovitz, *Atomic Diplomacy, Hiroshima and Potsdam* (New York, 1965); G. Kolko, *The Politics of War* (New York, 1968); D. S. Clemens, *Yalta* (New York, 1970); and L. C. Gardner, *Architects of Illusion* (Chicago, 1970).
50. FO 371/73048, Z 1345/273/72/G, 19 February 1948.
51. FO 371/73049, Z 1528/273/72/G, 18 February 1948.
52. FO 371/73050, Z 1764/273/72/G, 1 March 1948.
53. Bullock, op. cit., p. 526.
54. Ibid., p. 528.
55. *FRUS*, Vol. 3, 1948, pp. 32–3.
56. FO 371/73051, Z 1934/273/72/G, 7 March 1948.
57. FO 371/73051, Z 1999/273/72/G, 9 March 1948 and FO 371/73052, Z 2163/273/72/G, 12 March 1948.
58. FO 371/73051, Z 2003/273/72/G, 9 March 1948.
59. Ibid.
60. FO 371/73051, Z 2023/273/72/G, 10 March 1948.
61. Henderson, op. cit., p. 12.
62. FO 371/73055, Z 2559/273/72/G, 17 March 1948.
63. Henderson, op. cit., p. 14.
64. Barker, op. cit.
65. Ibid., p. 127.

66. CAB 128/14, CM (48) 19th Concls, 5 March 1948.
67. G. Warner, 'Britain and Europe in 1948: the View from the Cabinet', J. Becker and F. Knipping (eds), *Power in Europe? Great Britain, France, Italy and Germany in a Post-war World, 1945–1950* (New York, 1986), p. 34.
68. FO 371/67673, Z 8461/G, 22 September 1947.
69. FO 800/465, 29 November 1947.
70. FO 371/67673, Z 8579/G, 27 September 1947.
71. FO 371/71766, UR 603, 27 February 1947.
72. FO 371/73057, Z 3412/G, 17 April 1947.
73. Warner, op. cit., p. 37.
74. HC Deb., 5th. 446/395.
75. FO 800/465, 26 September 1948.
76. Ibid.
77. Warner, op. cit., p. 37.
78. John Kent and John Young, op. cit.

6 THE CHIEFS OF STAFF AND THE CONTINENTAL COMMITMENT

1. CAB 129/7, CP (46) 65, 15 February 1946.
2. At the end of the war about nine million people were involved either directly or indirectly in the war effort. British armed forces in Europe, the Mediterranean and the Far East numbered around 5.1 million. The rest were engaged in producing equipment and supplies for the forces.
3. CAB 131/1, DO (46) 1st Mtg, 11 January 1946.
4. CAB 131/1, DO (46) 22nd Mtg, 19 July 1946.
5. CAB 131/1, DO (46) 27th Mtg, 16 November 1946.
6. Bullock, op. cit., pp. 64–5.
7. CAB 119/159, CoS (44), 236 (0). See also CAB 79/71, CoS (44) 31st Mtg, Minute 3.
8. See J. Albert, *Attlee, the CoS and the Restructuring of Commonwealth Defence between V. J. Day and the Outbreak of the Korean War* (PhD, Merton College, 1986), p. 22.
9. CAB 79/45, CoS (46) 33rd Mtg (5), 1 March 1946.
10. Lewis, op. cit., p. 265.
11. Ibid., p. 267.
12. DEFE 4/4, CoS (47) 74th Mtg, Minute 1, 'The Overall Strategic Plan', withheld, 11 June 1947. See Lewis, op. cit., p. 97. See also Appendix 2.
13. CAB 138/88, PMM (48) 1st–15th Mtgs, 11 to 22 November 1948.
14. Albert, op. cit.
15. Albert argues that the amount of cooperation which occurred has been neglected by most writers on British defence policy in the immediate postwar period. He does, however, concede that the structure of Commonwealth defence established between July 1949 and June 1950 was 'a rickety edifice'.
16. CAB 80/99, CoS (46) 45 (0), 13 February 1946.
17. CAB 131/3, DO (46) 80. See Albert, op. cit., p. 76.

18. CAB 80/100, CoS (46) 54 (0), 22 February 1946.
19. Barker op. cit., p. 49.
20. CAB 133/86, PMM (46) 1, 20 April 1946.
21. Ibid.
22. See Barker, op. cit., p. 49. See also Field-Marshal Lord Montgomery, *Memoirs* (London, 1958), p. 436.
23. CAB 131/12, DO (46) 40, 13 March 1946.
24. CAB 131/1, DO (46) 10th Mtg, 5 April 1946; CAB 131/2, DO (46) 47, 2 April 1946.
25. CAB 133/86, PMM (46) 1, 20 April 1946.
26. Barker, op. cit., p. 53.
27. See J. Baylis, *Anglo-American Defence Relations, 1939–84, The Special Relationship* (London, 1983).
28. See R. Ovendale, *The English-speaking Alliance: Britain, the United States, the Dominions and the Cold War,* op. cit., pp. 45–6.
29. CAB 131/1, DO (46) 10th Mtg, 5 April 1946.
30. DEFE 5/3, CoS (47) 5 (0), 23 January 1947.
31. DEFE 5/5, CoS (47), 142 (0), 9 July 1947.
32. Albert, op. cit., p. 137.
33. DEFE 4/4, CoS (47) 74th Mtg, withheld, 11 June 1947. See J. Albert, op. cit., p. 123.
34. Ibid.
35. This was reflected in the Washington Talks on the Middle East in October 1947 which initiated a new era of Anglo-American defence cooperation in peacetime. Albert, op. cit., p. 137.
36. Montgomery, op. cit., pp. 435–6, 498.
37. CAB 79/54, CoS (46) 187 Mtg (1), 23 December 1946.
38. Ibid.
39. Ibid.
40. Lewis, op. cit., pp. 293–315.
41. DEFE 4/4, CoS (47) 74th Mtg, withheld, 11 June 1947.
42. See p. 77.
43. DEFE 4/4, CoS (47) 71st Mtg, 6 June 1947.
44. Barker, op. cit., p. 113. For a discussion of French interest in military staff talks and the visit of General Revers to Britain in January 1948 see John Kent and John Young, 'The Western Union Concept and British Defence Planning 1947–48', in R. Aldrich (ed.), *British Intelligence and Security Policy* (London, 1991).
45. FO 371/73045/273, 9 January 1948.
46. CAB 131/5, DO (48) 2nd Mtg, 8 January 1948.
47. Ibid.
48. DEFE 4/4, CoS (47) 16rd, 23 December 1947 and CAB 131/5, DO (48) 2nd, 8 January 1948.
49. DEFE 4/10, JP (48) 16 (Final), 27 January 1948.
50. Montgomery, op. cit., pp. 498 ff.
51. DEFE 4/10, CoS (48) 16th Mtg, 2 February 1948.
52. Ibid.
53. DEFE 4/10, CoS (48) 18th Mtg, 4 February 1948.
54. Ibid.

55. Ibid.
56. Ibid.
57. Ibid.
58. See Appendix 2.
59. See John Kent and John Young, op. cit.
60. CAB 128/12 CM (48) 19th, 5 March 1948.
61. John Kent and John Young, op. cit.
62. DEFE 4/11, CoS (48) 42nd Mtg, 19 March 1948.
63. DEFE 4/2, JP (48) 35 (Final), 30 March 1948.
64. A late revised version of the British plan was known as Speedway and the revised US plan was known as Fleetwood.
65. DEFE 4/13, CoS (498) 64th Mtg, 10 May 1948.
66. DEF 4/13, CoS (48) 66th Mtg, 12 May 1948.
67. Albert agrees that the stimulus of combined planning with the United States forced Britain to reconsider its attitude to the defence of Western Europe. Op. cit., pp. 137–41.
68. CAB 131/7, DO (49) 45, 17 June 1949.
69. CAB 13/18, DO (49) 16th Mtg, 21 June 1949.
70. DEFE 4/21, CoS (49) 75th Mtg, Minute 1.
71. Ibid.
72. Albert, op. cit., p. 325.
73. The Soviet atomic test appears to have had a major impact on British thinking. Britain could not afford to have the Soviet Union in control of the Channel ports. DEFE 4/29, CoS (50), 37th Mtg, Minute 8 and CAB 131/8, DO (50) 5th Mtg, Minute 1.
74. CAB 130/9, DO (50) 45 has been withheld. Information about the paper is contained in DEFE 5/20, CoS (50) 139, 28 April 1950. This document, however, was also recalled by the MoD in February 1986.
75. DEFE 4/131, CoS (50) 74th Mtg, 11 May 1950. Slim summarises CoS (50) 139 for Shinwell in this paper. Slim played a very significant part in changing the thinking of the other Chiefs of Staff towards a continental commitment from April 1949 onwards. The significant move in this direction, however, only came after the North Atlantic Treaty, binding the US to European defence, was signed.
76. The Soviet atomic test also played a significant part in this shift in British strategic planning.

7 THE PENTAGON TALKS, 22 MARCH – 1 APRIL 1948

1. DoS Files, Record Group (RG) 59, Box C-509, 840.20/3–1148, 11 March 1948.
2. Lord Gladwyn's review of Escott Reid, *Time of Fear and Hope*, op. cit., in *International Journal*, Vol. 33, no. 1, Winter 1977/8, p. 21. See also *FRUS*, 1948, Vol. III, pp. 64–7, 70–5.
3. A. Foster, 'The British Press and the Coming of the Cold War', op. cit.
4. Lord Gladwyn, *The Memoirs of Lord Gladwyn*, op. cit., p. 215.
5. See pp. 103–5, 112–15.

6. For an excellent account of the talks, see C. Wiebes and B. Zeeman. 'The Pentagon negotiations March 1948: the launching of the North Atlantic Treaty', *International Affairs*, Vol. 59, no. 3, Summer 1983. This chapter owes a great debt to the Wiebes and Zeeman article.
7. FO 371/68067, AN 1196/1195/45/G, 15 March 1948.
8. Public Archives of Canada (henceforth PAC), *Reid Papers*, Vol. 6, File 12, 4 November 1947. See also DoS File, 840.00/3–2348.
9. PAC, Reid Papers, Vol. 6, File 12, 18 March 1948.
10. Wiebes and Zeeman, op. cit., p. 355.
11. DoS Files, RG 59, Box C-507, 840.00/3–1948, 19 March 1948.
12. Wiebes and Zeeman, op. cit., p. 355.
13. G. F. Kennan, *Memoirs 1925–1950* (Boston, 1967), p. 409.
14. See Wiebes and Zeeman, op. cit., pp. 357–61. See also DoS Files, 840.00/3–2448 and 840.00/3–348.
15. PAC, Reid Papers, Vol. 6, File 12, Various drafts 23 March 1948.
16. Ibid., Draft of the Pentagon Paper, 24 March 1948.
17. FO 371/68067, AN 1326/1195/459, 25 March 1948.
18. FO 371/68068A, AN 1411/1195/45/G, 31 March 1948.
19. *FRUS*, Vol. III, 1948, op. cit., p. 74. See Appendix 4.
20. FO 371/68067, AN 1315/1195/45/G, 24 March 1948.
21. PAC, *Reid Papers*, Vol. 6, File 12, 23 March 1948.
22. *FRUS*, Vol. III, 1948, op. cit., p. 74. Emphasis added.
23. PAC, *Reid Papers*, Vol. 6, File 12, 27 March 1948.
24. *FRUS*, Vol. III, 1948, op. cit., p. 74. This included Spitzbergen and other Norwegian Islands, Iceland, Greenland, Newfoundland and Alaska.
25. FO 371/68067, AN 1315/1195/45/G, 25 March 1948.
26. Ibid.
27. PAC, *Reid Papers*, Vol. 6, File 12, 1 June 1948.
28. See Wiebes and Zeeman, op. cit., p. 363.
29. See pp. 105–6.
30. Henderson, op. cit., p. 33.

8 THE WASHINGTON TALKS ON SECURITY, 6 JULY 1948 TO 9 SEPTEMBER 1948

1. PREM 8/788, PM/48/38, 6 April 1948. See also FO 371/73069, Z 3941/2307/72/G, 7 May 1948.
2. FO 371/680A, AN 1431/1195/45/G, 1 April 1948.
3. FO 371/73069, Z 3941/2307/72/G, Z 4187/2307/72/G and Z 4188/72/G. See also PAC *Reid Papers*, Vol. 6, File 12, 8 May 1948.
4. *FRUS*, Vol. III, 1948, op. cit., p. 141.
5. Henderson, op. cit., p. 21.
6. Ibid., p. 22.
7. Ibid.
8. Ibid., pp. 23–4.
9. Ibid.

10. Ibid.
11. D. Cook, op. cit., pp. 157–8.
12. *FRUS*, Vol. III, 1948, op. cit., pp. 135–6.
13. D. Cook, op. cit., p. 83.
14. Henderson, op. cit., p. 25.
15. FO 371/73069, Z 3941/2307/72/G, 14 May 1985. Bevin's attempt to move the Americans forward received strong support from Louis St Laurent, the Canadian Secretary of State for External Affairs. In a speech to the Canadian House of Commons on 29 April he stressed the importance of the United States and Canada joining the Europeans in a security system within the United Nations Organisation.
16. FO 371/73070, Z 4269/2307/72/G. The Canadians also played a very influential role in defusing resistance from Kennan and Bohlen. See also George F. Kennan, *Memoirs 1925–1950* (London, 1967), Chapter 17 and C. Bohlen, *Witness to History* (London, 1973, p. 267, Bohlen strangely plays down his opposition to NATO. He says 'NATO was simply a necessity. The developing situation . . . demanded the participation of the United States.'
17. FO 371/73070, Z 4394/2307/72/G.
18. FO 371/73070, Z 5024/2307/72/G.
19. Franks felt that a personal meeting with Marshall was the only way to break the deadlock and lift the situation 'out of the groove'.
20. D. Cook, op. cit., p. 155.
21. Henderson, op. cit., p. 33.
22. FO 371/73072, Z 5362/2307/72 and Z 5454/2307/72/G. See *FRUS*, Vol. III, 1948, pp. 148–351. See also DoS Files, 840.20/7–648 and 840.20/7–748.
23. Henderson, op. cit., p. 36.
24. FO 371/73072, Z 5616/2307/72/G, 6 July 1948.
25. Henderson, op. cit., p. 41. See also FO 371/73074, Z 6140/2307/72/G.
26. Henderson, op. cit., p. 44.
27. FO 371/73077, Z 7563/2307/72/G and Z 7564/2307/72/G. See also FO 371/73073, Z 5819/2307/72/G.
28. Henderson, op. cit., p. 44.
29. Reid, op. cit., p. 45.
30. Ibid., p. 52.
31. Ibid.
32. FO 371/73072, Z 5454/2307/72/G.
33. Henderson, op. cit., p. 46.
34. Ibid., p. 52.
35. FO 371/73075, Z 6947/2307/72/G and FO 371/73077/, Z 7563/2307/72/G.
36. Henderson, op. cit., pp. 52–3.
37. Ibid.
38. See pp. 92 and 97.
39. Reid, op. cit., pp. 117–18.
40. Ibid.
41. Ibid.
42. FO 371/73075, Z 6947/2307/72/G.
43. FO 371/73077, Z 7589/2307/72/G.
44. FO 371/73077, Z 7564/2307/72/G.

9 LAST-MINUTE PROBLEMS, 9 SEPTEMBER 1948–28 MARCH 1949

1. FO 371/73081, Z 10182/2307/72/G.
2. PPS 43: 'Considerations affecting the conclusion of a North Atlantic Security Pact', *FRUS*, 1948, Vol. III, pp. 283–9. See also FO 800/483/NA/48/4.
3. Ibid.
4. See pp. 94–5.
5. Cook, op. cit., pp. 130–2.
6. Reid, op. cit., p. 148.
7. Henderson, op. cit., pp. 69–75. Reid, op. cit., p. 148.
8. FO 371/79224, Z 1161/1074/72/G.
9. FO 371/79225, Z 1416/1074/72/G.
10. FO 371/79226, Z 1465/1074/72/G and Z 1584/1074/72/G.
11. Henderson, op. cit., p. 91.
12. Reid, op. cit., pp. 149–50.
13. DoS Files, 840.20/2/–849.
14. Ibid.
15. See Henderson, op. cit., pp. 91–2.
16. FO 371/79226, Z 1584/1074/72/G, 17 February 1949 (emphasis added).
17. Ibid.
18. FO 371/79229, Z 1775/1074/72/G and Z 1778/1074/72/G. See also DoS Files, 840.20/2–1449.
19. D. Cook, op. cit., p. 214.
20. Henderson, op. cit., p. 98.
21. See Wiebes and Zeeman, op. cit., pp. 357–61.
22. DoS Files, 840.20/9–948.
23. Henderson, op. cit., p. 98.
24. 'North Atlantic Security Pact' (NASP), Canadian Archives, File 283(s), part 7.
25. Ibid.
26. See Appendix 5.
27. FO 371/79228, Z 1773/1074/72/G, 26 February 1949 and FO 371/79229, Z 1778/1074/72/G.
28. FO 371/79221, Z 283/1074/72/G and FO 371/79222, Z 519/1074/72/G, 18 January 1949.
29. Ibid.
30. FO 371/79229, Z 1870/1074/72/G, 1 March 1949.
31. FO 371/79228, Z 1733/1974/72/G, 26 February 1949.
32. *FRUS*, 1949, Vol. IV, pp. 126–39.
33. Reid, op. cit., p. 215.
34. Cook, op. cit., pp. 217–18.
35. This view is supported by Escott Reid, ibid., p. 216.
36. NASP, File 283(s), part 6.
37. *FRUS*, 1949, Vol. IV, pp. 129–31.
38. Ibid., pp. 206–12.
39. Cook, op. cit., pp. 218–19 and Henderson, op. cit., pp. 95–6.
40. FO 371/79232, Z 2081/1074/72/G.

41. Ibid., 9 March 1949.
42. See D. Cook, op. cit., p. 221.

10 CONCLUSIONS AND ACHIEVEMENTS

1. See pp. 20–7 and 27–30.
2. FO 371/67670/25, 21 December 1946. Quoted in Rothwell, op. cit., p. 435.
3. This is hardly surprising given Britain's severe economic difficulties in the immediate postwar period.
4. FO 371/66546. Quoted in A. Adamthwaite, 'Britain and the World, 1945–1949: the view from the Foreign Office', in J. Becker and F. Knipping, op. cit., p. 15.
5. See Bullock, op. cit., p. 520.
6. CAB 131/5, DO (48) 2nd Mtg, 8 January 1948.
7. Quoted in J. Kent and J. W. Young, '"The Western Union" concept and British defence planning 1947–1948', op. cit.
8. Barker, op. cit., p. 145.
9. FO 371/76384 and FO 371/76385. See also A. Adamthwaite, op. cit., p. 17.
10. Ibid.
11. Ibid.
12. Ibid.
13. Ibid.
14. The Soviet atomic explosion also played a key role in this decision.
15. CAB 139/9, DO (50) 45, withheld.
16. See Albert, op. cit.
17. FO 800/448/CONF/49/3. See Bullock, op. cit., p. 673.
18. Ibid.
19. Bullock, op. cit., p. 839.
20. Quoted in ibid., p. 840.
21. John Kent and John Young, op. cit.
22. This is a view which is also characteristic of some of this author's own earlier work. See John Baylis, *Britain and the Formation of NATO: A Study of Diplomatic Vision, Pragmatism and Patience*, International Politics Research Papers, no. 7, University College of Wales, Aberystwyth, 1989.
23. For a detailed survey of the opposition to Bevin's policies within the Labour Party see ibid., pp. 61–70.
24. Ibid., p. 846.
25. The relationship between Bevin and Byrnes was particularly turbulent at times.
26. It is perhaps difficult to refute that Bevin's attempt to keep his options open by pursuing close relations with the United States and Western European states in the security field undermined the chances of coming to an agreement with the Soviet Union. This had been the dilemma towards the end of the war. Bevin would have been irresponsible, however, not to have kept open other options in case relations with the Soviet Union did not work out satisfactorily. Contemporary relations were not promising.

27. Bullock, op. cit., pp. 62–3.
28. Uncritical assessments of Bevin include Alan Bullock, op. cit. and Sir Roderick Barclay, *Ernest Bevin and the Foreign Office, 1932–69* (London, 1975). For a more critical account see John Kent, 'The British Empire and the origins of the cold war', in A. Deighton (ed.), *Britain and the First Cold War* (London, 1990).
29. D. Cook, op. cit., p. 89.
30. A. Bullock, op. cit., p. 848.

Select Bibliography

Anderson, T. H., *The United States, Great Britain and the Cold War 1944–1947* (Columbia, 1981).

Attlee, C. R., *As it Happened* (London, 1954).

Barclay, R. E., *Ernest Bevin and the Foreign Office* (London, 1975).

Barker, E., *The British between the Superpowers 1945–50* (London, 1983).

Becker, J. and Knipping, F. (ed.), *Power in Europe? Great Britain, France, Italy and Germany in a Post-war World 1945–1950* (New York, 1986).

Best, R. A., *'Co-operation with Like-Minded Peoples' British Influences on American Security Policy, 1945–1949* (New York, 1986).

Bond, B., *British Military Policy Between the Two World Wars* (Oxford, 1980).

Bullock, A., *Ernest Bevin, Foreign Secretary 1945–1951* (London, 1983).

Cook, D., *Forging the Alliance, NATO, 1945–1950* (London, 1989).

Dilks, D. (ed.), *Retreat from Power: Studies in Britain's Foreign Policy in the Twentieth Century* (London, 1981).

Dixon, P., *Double Diploma: the Life of Sir Pierson Dixon, Don and Diplomat* (London, 1968).

Eden, D., *The Reckoning* (London, 1965).

Fitzsimons, M. A., *The Foreign Policy of the British Labour Government, 1945–1951* (Notre Dame, 1953).

Gladwyn, H. M. G. J., *The Memoirs of Lord Gladwyn* (London, 1972).

Gowing, M. M., *Independence and Deterrence: Britain and Atomic Energy, 1945–1952*, 2 vols (London, 1974).

Harris, K., *Attlee* (London, 1982).

Hathaway, R. M., *Ambiguous Partnership, Britain and America 1944–1947* (New York, 1981).

Henderson, N., *The Birth of NATO* (London, 1982).

Howard, M., *The Continental Commitment: The Dilemma of British Policy in the Era of Two World Wars* (London, 1972).

Ireland, T. P., *Creating the Entangling Alliance: The Origins of the North Atlantic Treaty Organization* (London, 1981).

Ismay, Lord, *The Memoirs of Lord Ismay* (London, 1960).

Jones, B., *The Russia Complex: the British Labour Party and the Soviet Union* (Manchester, 1977).

Kaplan, L. S., *The United States and NATO, The Formative Years* (Lexington, 1984).

Kennedy, P. M., *The Realities Behind Diplomacy: Background Influences on British External Policy, 1865–1980* (London, 1981).

Lewis, J., *Changing Direction: British Military Planning for Post-war Strategic Defence, 1942–47* (London, 1988).

Louis, W. R., *The British Empire in the Middle East 1945–51, Arab Nationalism, the United States, and Post-war Imperialism* (Oxford, 1984).

Macmillan, H., *Tides of Fortune* (London, 1969).

McNeill, W. H., *America, Britain and Russia. Their Co-operation and Conflict, 1941–1946* (London, 1953).

Manderson-Jones, R. B., *The Special Relationship: Anglo American Relations and Western European Unity 1947–1956* (London, 1972).
Medlicott, W. N., *British Foreign Policy since Versailles 1919–1963*, (London, 1968).
Monroe, E., *Britain's Moment in the Middle East 1914–1971* (London, 1981).
Montgomery, B. L., *The Memoirs of Field-Marshal the Viscount Montgomery of Alamein* (London, 1958).
Moran, C. M. W., *Winston Churchill, The Struggle for Survival, 1940–1965* (London, 1966).
Nicholas, H. G., *Britain and the United States* (London, 1963).
Northedge, F. S., *Descent from Power: British Foreign Policy 1945–1973* (London, 1974).
Ovendale, R. (ed.), *The Foreign Policy of the British Labour Governments, 1945–51* (Leicester, 1984).
Ovendale, R., *The English-Speaking Alliance: Britain, the United States, the Dominions and the Cold War 1945–51* (London, 1985).
Pearson, L. B., *Mike. The Memoirs of the Rt. Hon. Lester B. Pearson, Vol. 1, 1897–1948* (Toronto, 1972).
Pearson, L. B., edited by Munro, J. A., and Inglis, A. I., *Mike. The Memoirs of the Rt. Hon. Lester B. Pearson, Vol. 2, 1948–1957* (Toronto, 1973).
Reid, E., *Time of Fear and Hope: The Making of the North Atlantic Treaty, 1947–1949* (Toronto, 1977).
Rendel, Sir G., *The Sword and the Olive, Recollections of Diplomacy and the Foreign Service, 1913–1954* (London, 1957).
Rothwell, V., *Britain and the Cold War, 1941–1947* (London, 1982).
Rubin, B., *The Great Powers in the Middle East 1941–1947, The Road to the Cold War* (London, 1980).
Shinwell, E., *Lead with the Left: My First Ninety-Six Years* (London, 1981).
Smith, F. E., *Halifax: The Life of Lord Halifax* (London, 1965).
Smith, J., *The Origins of NATO* (Exeter, 1990).
Staercke, A. de. (ed.), *NATO's Anxious Birth: The Prophetic Vision of the 1940's* (London, 1985).
Stephens, M., *Ernest Bevin: Unskilled Labourer and World Statesman 1881–1951* (London, 1981).
Strong, W., *Home and Abroad* (London, 1956).
Thorne, C., *Allies of a Kind: the United States, Britain and the War against Japan 1941–1945* (London, 1978).
Watt, D. C., *Succeeding John Bull, America in Britain's Place 1900–1975* (Cambridge, 1984).
Wheeler-Bennet, J., and Nicholls, A. *The Semblance of Peace: The Political Settlement after the Second World War* (London, 1972).
Williams, F., *Ernest Bevin, Portrait of a Great Englishman* (London, 1952).
Williams, F., *A Prime Minister Remembers: The War and Post-war Memoirs of the Rt. Hon. Earl Attlee* (London, 1961).
Woodward, Sir L., *British Foreign Policy in the Second World War, Vol. V.* (London, 1976).
Young, J. W., *Britain, France and the Unity of Europe, 1945–51* (Leicester, 1984).

Index